REALISM AND REALITY
The Novel and Society in India

REALISM AND REALITY
The Novel and Society in India

MEENAKSHI MUKHERJEE

DELHI
OXFORD UNIVERSITY PRESS
BOMBAY CALCUTTA MADRAS
1994

Oxford University Press, Walton Street, Oxford OX2 6DP
OXFORD   NEW YORK   TORONTO
DELHI   BOMBAY   CALCUTTA   MADRAS   KARACHI
KUALA LUMPUR   SINGAPORE   HONG KONG   TOKYO
NAIROBI   DAR ES SALAAM   CAPE TOWN
MELBOURNE   AUCKLAND   MADRID
and associates in
BERLIN   IBADAN

ISBN 0 19 563434 9

Printed at Rekha Printers Pvt. Ltd., New Delhi 110020
and published by Neil O'Brien, Oxford University Press
YMCA Library Building, Jai Singh Road, New Delhi 110001

For Bapi and Maiya

ACKNOWLEDGEMENTS

Sections from Chapters II, III, IV and VI have appeared in earlier versions in the following journals:

Journal of Commonwealth Literature, August 1981
Economic and Political Weekly, 29 May 1982 and 14 January 1984
Journal of Arts and Ideas, August–October 1983

PREFACE TO THE FIRST EDITION

This book has been a long time in the making. Many years ago while writing my doctoral dissertation on Indian novels in English I made a cursory exploration into the Indian novels available in translation in order to make an incidental point in my opening chapter. It soon became evident that the published material, even in translation, was too vast and too rich to be dealt with casually and relegated to a footnote. One obviously needed several years to study these novels, relate them to their historical and social contexts, and construct a theoretical framework along which the development of the novel in India could be aligned. What began for me as incidental curiosity grew gradually into a major concern that has persisted.

The book could not be completed earlier because I was able to concentrate on it only intermittently and in short spells while teaching full-time. The usual fund-granting authorities in the country who can and do help research workers to ride their hobby horses found my project rather unorthodox for their taste. This in a way strengthened my determination to carry on and prove at least to myself that what I wanted to do was a viable line of research—namely to take a holistic view of the novel in India as a genre nursed by if not born out of the tension between opposing systems of value in a colonial society, and modified by certain indigenous pressures. I do not know if I have been able to achieve even a fraction of what I set out to do, but in the attempt I have learnt a good deal. I found for example that the conventions of realism—the dominant mode of nineteenth-century British fiction and the immediate model for the first generation of the Indian novelists—could not be transferred to the Indian situation, where the nature of social reality was substantially different, without causing certain inadvertent mutations in the mode itself. The fundamental problem of the early novelists of India was how to reconcile the demands of realism with the intransigence of reality.

While working on this book I sometimes encountered queries about the feasibility of a project in literary criticism wherein the texts are in different languages. The general belief, even in our faculties of literature, is that while an isolated study of the Tamil novel or Indo-Anglian fiction can be a legitimate academic activity, any attempt to study the Indian novel in a broader perspective is bound to be a futile endeavour. This book is an attempt to refute that argument. It is my contention that the novel in India can be seen as the product of configurations in philosophical, aesthetic, economic and political forces in the larger life of the country. Despite obvious regional variations, a basic pattern seems to emerge from shared factors like the puranic heritage, hierarchical social structure, colonial education, disjunction of agrarian life and many others that affect the form of a novel as well as its content. The variables, however, sometimes outweigh the commonalities, hence the discussion has to steer clear of any schematic categorization.

In order to pre-empt criticism of the theoretical premises in the book I must stress what I have *not* tried to do. I have not tried to write a history of the novel in India; that would require at least a dozen like-minded collaborators. Nor have I tried to make a comprehensive study of the genre in India because I have depended heavily on translations. My access to the originals was limited by the fact that I read only two Indian languages directly and a third with some help. Thus, among the novels discussed here I have read only the Bengali, Hindi and some of the Marathi texts in the original. For the rest I have had to use either English renderings or translations into Hindi or Bengali. The 'Adaan-Pradaan' series of the Book Trust (which arranges to get major Indian novels from different languages translated into different Indian languages) has proved to be an invaluable quarry for my purpose. It is risky, all the same, to talk about trends and patterns when one is working with translated material because what gets translated and what does not depends on various chance and extraneous factors. The quality of a translation can also influence the critic's judgement unconsciously. These are severe limitations that I have kept in mind throughout and, consequently, desisted from making broad claims.

The book is divided into two parts, the first indicating the possibility of conceptual frames, the second trying to validate these concepts through textual analysis of individual novels. The co-ordinates along which the novel in India can be charted are drawn in the first chapter. In Chapters II, III and IV three strands are picked up out of the literary tangle of late-nineteenth-century India for closer study. Chapter V deals with a single novelist of the early-twentieth century in whom the popular fiction reading taste of all regions of the country converged.

In the second part, three works—*Pather Panchali* (1929), *Godan* (1936) and *Samskara* (1965)—are discussed in some detail, both in terms of form and theme as illustrations of the theoretical arguments offered in the earlier chapters. All three are major novels in their respective languages and they are also available in English translation to those who cannot read them in any other language.

Among the many people who have helped with books, information and ideas, I am specially grateful to Ashok R. Kelkar (Pune), K. Ayyappa Paniker (Trivandrum), Sisirkumar Das (Delhi), R. Srinivasan (Bombay), Nirmalya Acharya (Calcutta), Gopal Rai (Patna), Parthapratim Bandopadhyay (Naihati), Gail Minault (Austin), Rajen Harshe and Shashi Mudiraj (both from Hyderabad), my former students T. Vijay Kumar and Vasundhara Bhalla, and also my daughter Rohini, who has helped with the index. I am indebted to Dr S. Nagarajan and other colleagues at the University of Hyderabad who so readily concurred with the idea of my offering a course in the English Department entitled 'The Novel and Society', where I used Indian novels in English translation and derived the benefit of classroom discussion. My thanks are also due to Shri W. H. Patwardhan for goading me on to finish the work, to Dr F.T. Jannuzi of the University of Texas at Austin for inviting me to teach a course on the novel in India at the Centre for Asian Studies, where some of the ideas could be tested out; and last as well as foremost to Sujit Mukherjee for fruitful disagreements and agreeable fruits of argument.

Hyderabad
August 1984

Meenakshi Mukherjee

PREFACE TO THE PAPERBACK EDITION

This book was published nearly a decade ago and some of the chapters were written considerably earlier. But no attempt has been made to substantially change the text for the present edition because revision might have resulted in a new book altogether. Much has happened in the intervening years in the study of narrative genre as well as in the field of colonial discourse theory. The new readers of *Realism and Reality*, it is hoped, will take the arguments of this book further in the light of these new perspectives.

New Delhi
May 1993   Meenakshi Mukherjee

CONTENTS

PART ONE

I

FROM PURANA TO NUTANA

A study of the emergence of the novel in India has to be more than a purely literary exercise. The factors that shaped the growth of this genre since the mid-nineteenth century arose as much from the political and social situation of a colonized country as from several indigenous though attenuated narrative traditions of an ancient culture that survived through constant mutation. English education and through English an exposure to western literature were by far the strongest influences at work. It is not an accident that the first crop of novels in India, in Bengali and Marathi, appeared exactly a generation after Macaulay's Educational Minutes making English a necessary part of an educated Indian's mental make-up were passed. Yet to regard the novel in India, as is sometimes done,[1] as purely a legacy of British rule—such as the railways or cricket—would be to overlook the complex cultural determinants of a literary genre.

When the novel assumed a distinct generic identity in Europe in the eighteenth century its form was quite different from that of the existing structures of earlier narrative such as the epic, the romance or the saga. Since the novel is the second-youngest major narrative genre today (the film being the youngest), there has been considerable critical and philosophical speculation in the West about why it emerged when it did and whether there was a historical inevitability about its emergence. The co-ordinates taken into consideration by critics as different as Hegel, Lukacs, Steiner, Watt and Todorov include economic and political factors as well as metaphysical assumptions about man's relationship with time, with nature and with other human beings. Formal realism, regarded as one of the determining characteristics of the novel in its early stage, reflected a basic shift in man's view of reality. Of the many theories about the rise of the novel in the West the two major theories emphasize the novel's close connection with the changing economic and moral bases of society and, as we shall see, its members' awareness of the temporal and spatial axes of reality.

Stripped to its essence, one theory may be said to attribute the rise of this genre to the growth of the bourgeoise and modern capitalism.[2] That is, the novel is said to be a genre generated and sustained by the middle class in a very broad sense, and to incorporate the values of this class as against the feudal values of the epic or romance. While the epic hero's fate was linked with the destiny of his community and the romance hero followed a predetermined heroic or chivalric code, the protagonist of the novel has to make a choice which is distinctly his own. The second theory suggests a link between the emergence of the idea of individualism and the rise of the novel.[3] But this concept of individualism can also be related to the new social mobility that industrialization made possible, displacing man from his secure traditional niche, making him realize the unique potential of each human being, including himself, outside social hierarchy. These two theories are thus not unrelated.

All this is evident to any student of the novel in the West. But how much of this can be transferred to a different cultural and historical context when we study the novel in India? Is it possible to say that industrialization or the rise of the middle class need not be a necessary precondition for the new form because Banabhatta's *Kadambari* was written in Sanskrit in the seventh century and because tenth-century Japan produced the long narrative in prose fiction well known in English translations as *The Tales of Genji? Don Quixote*, a much more direct ancestor of the European novel, was written in 1605 in a pre-industrial Spain steeped in feudal values. The crux of the problem may lie in deciding upon a set of defining characteristics of the novel which are valid across cultures. However, in the two-and-a-half centuries or more that have elapsed since the novel was recognized as a distinct genre, it has become increasingly clear that this is the most flexible and elusive of genres, almost impossible to tackle through definitions. The most one can attempt is to describe some of the obvious differences between this genre and earlier forms of narrative.

In pre-novel narratives, for example, *Kadambari, Panchatantra, Arabian Nights, The Tales of Genji, Legends of King Arthur, Decameron* or *Canterbury Tales*, the narrative structure is often circular—i.e. either there is a larger story which contains a smaller one which in turn contains another and so on, or a number of shorter tales are strung together in the larger thread of the central narrative. In *Kadambari* Shudraka listens to the account of the Suka bird; the Suka bird listens to Maharishi Jabali's story which contains Mahashveta's story within it. Such cycles or chain tales have existed in almost every language, though not always in prose, and have been variously called sagas or romances. The Persian word *daastan*, later carried over to Urdu, denotes similar cycles of medieval tales of heroism where a succession of episodes follow one another in endless profusion. Compared to these the structure of the novel is more or less unified. The events grow organically out of each other instead of being loosely strung together through a common thread. On the whole the progression is linear rather than cyclic, even though the order of past and present may occasionally be reversed.

The consciousness of time and space is a special feature of the novelist's apprehension of reality. The pre-novel tales have a 'once-upon-a time' ambience where the tensions of time past and time present are absent. Instead of dealing with the unchanging moral verities of life in the abstract, the novelist depicts situations on spatial and temporal axes, employing realism as one of the viable modes of viewing this concrete human reality. Medieval tales could be borrowed by one culture from another (Chaucer's *Pardoner's Tale*, for example, can be traced to the Jataka stories) but a novel is necessarily bound by its historical and geographical co-ordinates. An organic product of a specific environment in a particular society at a given point of history, the novel crosses the frontiers of culture less easily than a fable or an allegory.

The third important distinction between the novel on the one hand and fable, legend and all other traditional narratives on the other lies in characterization, which is 'life-like' in the one and stylized in the other. The archetypes of the hero, the heroine and the villain of the romance have been seen by Northrop Frye as the reflections of Jung's libido, anima and shadow respectively. Frye says that 'a suggestion of allegory is constantly creeping in around its [the romance's] fringes',[4] whereas in a novel the characters are seen not as representatives of either a class or of moral values but as specific individuals who are required to be convincing in the context of a given time and culture.

II

A whole new world became available to educated Indians in the middle of the nineteenth century through their study of English literature. The society represented in the novels of Scott, Dickens and Thackeray was very different from the society Indians knew and lived in, which in turn was already different from the traditional agrarian life of previous generations which had been the stable cultural background — a background which the extension of British rule had partially disrupted. Since the early novels in India were all written in urban areas by English-educated people (this remains true even today), this discontinuity was indeed a vital issue. It was responsible for their inability to find a fictional form suitable for the new city society, a form which simultaneously allowed novelists to employ, without doing violence to the reality of mid-nineteenth century 'Indian life', the realist mode learnt from the Victorian novel. More influential than Dickens and Thackeray were popular Victorian novelists like Wilkie Collins, Marie Corelli, Benjamin Disraeli, Bulwer Lytton, and a now-forgotten manufacturer of bestsellers called G.W.M. Reynolds. Moth-eaten copies of books by these writers can be found in almost every family library in India which goes back to the nineteenth century. Not all these British novelists used the realist mode, but for the nineteenth-century Indian it was not very easy to distinguish precisely between fictional modes when the life depicted was so unfamiliar. Colourful, expansive, free—the characters in the work of these novelists seemed to lead lives of infinite possibilities, while the life of the nineteenth-century Indian—politically servile, economically deprived and socially circumscribed—seemed to them limited in comparison.

The picaresque tradition in the European novel had achieved one main purpose—it had liberated the protagonist from the rigidity of a static society into being a free agent who could to some extent shape his own destiny. *Robinson Crusoe* (1719), *Moll Flanders* (1722), *Pamela* (1740), three early examples of the English novel, show how the central character is in each case an active rather than a passive agent challenging his or her fate. The Indian novelist had to operate in a tradition-bound society where neither a man's profession nor his marriage was his personal affair. His life was mapped out by his family or his community or his caste. In the rigidly hierarchical familial and social structure of nineteenth-century India, individualism was not an easy quality to render in literature. One of the problems of the early novelist was to reconcile two sets of values—one obtained by reading an alien literature and the other available in life. It may be relevant here to quote two passages where two nineteenth-century novelists try to rationalize their predicaments. The first passage is from the introduction to a Marathi novel, *Manjughosha* (1868), written by Naro Sadashiv Risbud, who opts out of the realist mode altogether:

Because of our attitude to marriage, and for several other reasons, one finds in the lives of us Hindus neither interesting vices nor virtues, and this is the difficulty which we find in trying to write novels. If we write about the things we experience daily, there would be nothing enthralling about them, so that if we set out to write an interesting book we are forced to take up with the marvellous . . .[5]

The second passage is taken from the dedication of O. Chandu Menon's Malayalam novel, *Indulekha* (1888), where there is a more direct reference to the essential hurdle—writing in a form that requires individualism as a value and writing about a society that denies it:

As stated at the outset, my object is to write a novel after the English fashion, and it is evident that no ordinary Malayalie lady can fill the role of the heroine in such a story. My Indulekha is not, therefore, an ordinary Malayalie lady.[6]

This author however goes on to add that if an Indulekha ever became possible she would be found among the educated Nair women of Malabar who enjoyed (perhaps because of matri-lineal property laws) more freedom than most Indian women. The novel thus projects into the future rather than reflects a society known to the author:

Twenty years hence there may be found hundreds of Indulekhas in Malabar who would be able to choose their husbands for pure and sweet love. My narrative of the love and courtship of Madhavan is intended to show to the young ladies of Malabar how happy they can be if they can have the freedom to choose their partners.[7]

Indulekha actually ended up doing a lot more than what the author naively proposed in his preface and dedication. Starting as a professed adaptation of a second-rate Victorian novel (Disraeli's *Henrietta Temple*), it turned out to be the first major novel in Malayalam. This text is a good example of how despite self-perceived shortcomings and a rhetoric of humility a writer can trascend his limited model through a firm grasp of the milieu and time to which he belongs.

But the man–woman relationship, one of the staples of the European novel, presented the most persistent obstacle to the Indian writer who lived in a society bound by extremely restrictive conventions of marriage. Where girls were married off by their parents before puberty and marriage was a social institution rather than an act of individual choice, there was very little scope for romantic pre-marital love of the kind depicted in the English novels being read by the educated urban Indians. Love could be shown in an indigenous setting only in historical romances where the demands of realism were absent. The other alternative was a depiction of illicit love, but this involved the subsidiary problem of juxtaposing individual aspiration and the stability of the social order. In the novels of Bankimchandra Chatterji (1838–94) this conflict between rebellious passion and the accepted social norm sometimes becomes a central concern (for example in *Krishnakanter Will* and *Visha Vriksha*), but even at the cost of the artistic integrity of the novel Bankim had to accede that the demands of social order were higher. This was consistent with the other effect of English education, namely the desire to reform Hindu society or at least purge it of its excesses. There is otherwise enough evidence of free love between man and woman in ancient as well as in medieval India—as preserved in poetry or carved on stone.

A steadfast consciousness of the present is another pre-requisite of concrete characterization, because characters in a novel, unlike in myth or romance or epic, exist within a specific time. The awareness of history as an irrevocable process was a relatively recent phenomenon even in the nineteenth-century West. Ancient Greece had developed the spatial sense, and the past in the classical world was of value merely as an accretion of independent events which were complete in themselves. Post-Renaissance Europe began to see an organic quality in the process of history, with one state formed out of another. The development of the European novel coincided with the emergence of this dynamic view of time, and the structure of eighteenth and nineteenth-century European novels was indirectly based on the idea of a linear and sequential progression of events that happened along a temporal axis.

This brings us to the question of time, rather of the operation of time within the narrative structure. Mythic time is necessarily different from historic time. While the latter operates in a novel, the narrative structure of conventional *kavya* works reveals all time as part of a cosmic cycle. Not too much emphasis can however be given to this contrast in a study of the novel in India, because nineteenth-century Indian writers were influenced largely by western concepts. Their conscious models were Scott's and Thackeray's novels rather than *Brihatkatha* or *Kadambari, Dasakumaracharita* or *Kathasaritsagara*. Yet the unconscious influence of these works, of the puranic tradition, of oral narratives and the memory of episodes from *Ramayana* and *Mahabharata* on which the imagination of most Indian writers was sustained, cannot be ignored altogether.[8] A concept of time that does not put too high a premium on the progress of events or the uniqueness of each moment will in some way affect the novelist's apprehension of reality. It is possible to argue that realism, a characteristic technique of the novel from Defoe to Balzac, reflects a particular world-view at a certain phase in human history. The fidelity to actuality involves a focusing on the immediate, the here and now, on details of the visual world, on specific human action and its verifiable consequences. Indian literature did not have any tradition of this variety of realism because it was based on a rather different view of reality. Even when the nineteenth-century Indian writer started consciously to emulate the western writer, interest in the palpable surface of physical reality was slow to evolve. Descriptions of the sky or a sunset or a landscape are often found as stylized set pieces in these early novels as well as the elaborate *nakha-sikha* (literally, 'from toe nail to the top of the head') reports of the heroine's beauty in the Sanskrit kavya convention. But a realistic presentation of actual people or objects, interiors or buildings, was either absent or rare. Without going as far as V. S. Naipaul,[9] who theorizes that Indians are impervious to external details and who cites various examples to show how outer reality is used in Indian writing merely to preserve the continuity of the self, one can suggest that in the Indian literary tradition the perceptible surface of reality never had the same value as in Defoe or George Eliot, Balzac or Tolstoy. It is not surprising therefore that the two eighteenth-century novels most popular among the early generation English-educated Indians were *Rasselas* and *The Vicar of Wakefield*—both of which emphasized moral qualities rather than narrated amoral adventures in the realistic settings of Defoe and Fielding.

Thus the determinants of a literary form can be non-literary. Even religion can influence genre, as Edward Said has pointed out in a different context. Speculating on the absence of novels in the Arabic language until the present century he writes:

There was no tradition out of which these modern works developed; basically at some point writers in Arabic became aware of European novels and began to write like them. Obviously, it is not that simple.

Nevertheless, it is significant that a desire to create an alternative world, to modify or augment the real world through the act of writing (which is one motive underlying the novelistic tradition in the West) is inimical to the Islamic world-view. The Prophet is he who has *completed* a world-view; thus the word heresy in Arabic is synonymous with the verb 'to innovate', 'to begin'.[10]

The situation in India was quite different from the one suggested above, but Said's observation reinforces what has been argued so far—namely that unexpected extra-literary factors contribute to the emergence of a literary form or retard its development. Increasing interaction among cultures of the modern world has made it progressively difficult to attempt isolating such factors.[11] Even categorical distinctions between what is Indian and what is western in literature is fraught with danger. All that we can do is note the differences in given conditions, so that in the analysis of actual texts different literary consequences may seem less strange or inexplicable.

III

While conceding that the authentically modern writer in the twentieth century often chooses not to respect the separation between literary genres, Tzvetan Todorov has observed: 'It is because genres exist as an institution that they function as horizons of expectation for readers and as models of writing for authors.'[12] Europe's more stratified nineteenth-century thought is bound to have communicated its respect for genres to contemporary Indians who came in touch with them. But were those who wrote our early novels conscious of founding a new genre in Indian literature? Did their readers immediately recognize it as such and adjust their expectations accordingly? Neither the word 'novel' nor any of its Indian equivalents was applied to either *Yamuna Paryatan* (Marathi; 1857) or *Alaler Gharer Dulal* (Bengali; 1858), although in retrospect we can recognize in these works the beginnings of a new literary form. The word *upanyas*, now current in many north Indian languages as a synonym for 'novel', was first used in 1862 in Bengali by Bhudeb Mukhopadhyay when he named a volume containing two long tales set in the past as *Aitihasik Upanyas* (Bengali; 1862). The title could be translated as 'Historical Fiction', although the history contained in the tales does not claim to be verifiable. As for the word 'upanyas', which is of Sanskrit origin–meaning some statement properly presented or arranged in an orderly manner(even today in Telugu the word does not mean 'novel' but refers to a discourse or a speech)–it had never been used before to signify a long prose narrative. So when Bhudeb Mukhopadhyay employed the word he was obviously trying to coin a term for a new category of story. One wonders whether he was aware of the contemporary importance of his casual coinage–the tales he presented under this label do not appear to have been consciously modelled on European lines–and of its future utility. By the time Bankim began writing his novels in the mid-1860s, the term 'upanyas' was already well established in Bengali and about to be taken over in Hindi as well, as we note from the title *Manohar Upanyas* of a work of fiction published in 1871.[13]

In Marathi on the other hand the term which came to be used for the novel is *kadambari*, and the novelist is called a *kadambari-kar*. Such usage obviously pays tribute to Banabhatta's *Kadambari* and acknowledges it as the first literary work in this genre, an example of a name (such as xerox or frigidaire today) converted into a common noun. The use of the word 'kadambari' in this sense pre-dates the term 'upanyas' because a Marathi dictionary compiled in 1829 under the sponsorship of Montstuart Elphinstone lists 'kadambari' as meaning a fictional narrative in prose.[14] From Marathi the term passed on to Kannada.

Urdu resolved the matter differently. When Ratan Nath Sharshar published his *Fasana-i-Azad* (1868), he claimed that this work was something wholly unprecedented in Urdu fiction and he called it a *naval*. Since the English word 'novel' for this form implied newness as a feature of the form, the word 'naval' for the Urdu novel–with its approximation in sound as well as sense–was a happy choice. Gujarati extended this choice further by adapting the term *naval-katha*, which combined the element of newness with a reminder of tradition evoked by the

Sanskrit word 'katha' (meaning story). Tamil and Malayalam have borrowed the English term itself. So has Telugu, although early Telugu novels like Narahari Gopalakrishnaiah Setty's *Sri Rangaraja Charitram* (1872) and Kandukuri Veerasalingam's *Rajasekhara Charitram* (1878) were known as 'vachana prabandha', a loose translation of which could be 'prose fiction'.

These differences in terminology do not really matter because whatever term for the novel was adopted in an Indian language, the formal and thematic aspirations of the early Indian novel were the same as those of the English novels read by pioneering Indian novelists. The English-educated generation which came of age in India around 1860 was brought up on British Victorian novels of the time and seems to have been influenced by these. Although Lukacs has insisted that 'the primary determinants of such influences are the literary requirements of the recipient country',[15] sometimes the issue was determined by what English novels were actually available in India. The writers most often translated into Indian languages were Wilkie Collins, Disraeli and Reynolds among the Victorians, and Bunyan, Johnson and Goldsmith among the older writers. The popularity of some of these writers in India apparently continued even into the following century. Recounting his Bloomsbury experiences Mulk Raj Anand has recalled Virginia Woolf saying to him that she always thought the only popular writers were Galsworthy, Bennett and Wells, but her husband—who had been a civil servant in Ceylon—said he knew all the while that the really popular writers were W.M. Reynolds and Marie Corelli, 'the low-brow fodder...on which the subalterns chew their cud in cantonments of the empire'.[16]

The motivating impulse of the early writers of fiction in India varied greatly. The earliest long prose-narrative in Hindi in the nineteenth century, *Rani Ketaki ki Kahani* (1801) by Insha Allah Khan, was written as a linguistic experiment. The author wanted to show that a story could be written in a language which was neither Persianized Urdu nor a localized dialect of Hindi. The form was incidental, the language was the challenge. There was no European influence here, nor did the book generate any further experiments to begin a tradition.

In 1868 the Gujarati writer Nandshankar Tuljashankar Mehta wrote in the introduction to his historical romance, *Karana Ghelo*: 'The former education inspector of our State [Surat] Mr Russel has expressed to me his desire to see Gujarati books written along the lines of English novels and romances. I have written this novel according to that plan.'[17] He was not the only one who followed the initiative provided by a British official. The first Telugu novel was written as a response to Lord Mayo's announcement of a prize to a prose fiction 'depicting the customs and traditions of society'.[18] Many of these early works of fiction owed their origin and survival to official patronage. British officers often helped the authors by prescribing the works as textbooks, thereby ensuring sales, by awarding cash prizes or by arranging for bulk purchase. Pandit Gauri Dutt, the author of *Devrani Jethani ki Kahani* (Hindi; 1870), acknowledges with gratitude his debt to Mr M. Kempson, the Director of Public Instruction, who bought two hundred copies, and to the Lt. Governor who gave him a prize of Rs 100. British officials also helped by undertaking translations. The same Mr Kempson translated Nazir Ahmad's Urdu moral tale *Tabut-un-Nusuh* into English as *The Repentance of Nusooh* in 1884. Nazir Ahmad's earlier novel *Mirat-ul-Arus* (1869) was translated into English by a retired civil servant, G. E. Ward, as *The Bride's Mirror*. *Indulekha* (Malayalam; 1888) was translated into English by John W. F. Dumergue, a civil servant in the Madras Presidency.

In some cases the motivating impulse was simply to provide instruction and delight. Nazir Ahmad, in the preface to his first Urdu book, *Mirat-ul-Arus*, explains that he wrote it to provide his daughters with interesting reading material because they had nothing but sacred texts to read:

Purely religious subjects of study are not suited to the capacities of children, and the literature to which my children's attention was restricted had the effect of depressing their spirits, of checking their natural instincts and of blunting their intelligence ... It was then I formed the design of the present tale.[19]

Samuel Vedanayakam Pillai, in the preface to the first Tamil novel, *Prathapa Mudaliar Charitram* (1879), stated his didactic intentions clearly: 'My object in writing this work of fiction is to supply the want of prose works in Tamil... and also to give a practical illustration of the maxims of morality'.[20] Bankim did not discuss his intentions in the prefatory matter of any of his early novels. His narrative style and authorial intrusions testify to his having read Scott and Thackeray, but in sonorous passages of nature description and evocations of feminine beauty, conventions of Sanskrit literature are also evident. In almost every major novel of the nineteenth century, behind the obvious European influences can be found the bedrock of a different narrative structure and value systems. Chandu Menon, who proclaimed in the Introduction to *Indulekha* his desire to write in Malayalam a realistic novel 'in the English style', forgot the intention by the time he finished the story. The concluding lines echo the sentiments with which the oral recital of a purana traditionally ends: 'All the characters mentioned ... have reached the summit of human happiness, and now may God bless us and all who read this tale.'[21] The last line of this passage reveals the persistence of the pre-novel conventions of narrative in spite of the author's conscious adoption of the European mode and his deliberate debunking of the mythic imagination (see Appendix II).

On the kaleidoscopic fictional scene in nineteenth-century India it is not easy to impose any pattern. The novel did not develop at the same pace in every language, nor was the influence of English evident to the same degree. Considering the fact that the trading centres at Calcutta, Bombay and Madras had an earlier and greater exposure to western ways of life and thought than other parts of India and considering that the three universities in India were established simultaneously in these cities in 1858, at least in Bengali, Marathi and Tamil the develop-ment of modern literature should have followed similar lines. But in actual fact the variables were many more than the common factors, and these lay in the religious, cultural and political arenas. While differences in literary trends in different languages are considerable, certain common patterns also become perceptible if we allow for a time lag in the development of these patterns in different regions. There was a sudden spurt of long narrative fiction in most Indian languages in the second half of the nineteenth century, whether these were called upanyas, kadambari, naval-katha or novel, and at least three dominant strands can be sorted out from the tangled skein. The first strand consists of the novels of purpose which utilized this new literary form for social reform and missionary enterprise. The second is an inclusive category where the apparently opposed tendencies of historical and supernatural fiction merge, the common denominator being the creation of an ethos remote in time. The third strand attempted to render contemporary Indian society realistically in fiction, joining the European novelists 'in that effort, that willed tendency of art to approximate reality.'[22] This was perhaps the most important strand and it subsequently came to form the mainstream along which the Indian novel developed in the twentieth century, although the other two streams have never been invisible for too long. These streams are discussed separately in the three subsequent chapters and a few representative texts analysed from each, but in actual fact these often overlapped and merged.

IV

In recalling the difficulties faced by nineteenth-century Indian novelists in their attempt to adapt an imported form to suit indigenous requirements, an important point remains to be made: the late emergence of prose literature in almost all the modern Indian languages. Until fairly late in the eighteenth century, literature in India was almost synonymous with verse composition.

Pramatha Chaudhuri (1868–1948), a Bengali writer with a flair for succinct and witty phrasing, once said that when the British came, rhyme gave way to reason. The development of prose in many regions—at least around Calcutta, Bombay and Madras—happened at the initiative of Christian missionaries who set up printing presses to produce material in regional languages. Before this, verse had reigned supreme for centuries in most Indian compositions, not only in imaginative literature but also in astrological, medical, biographical and philosophical texts. Even until quite late, handbooks on homeopathy, texts of legal procedures and mathematical calculations were, in Bengali, available in verse.

Sunitikumar Chatterji regards the year 1800 as a pivotal one as far as the development of prose writing in most Indian languages is concerned. In Bengali, earlier prose was in the form of letters and documents. Therefore when the first Bengali novels emerged in the 1850s they were written in a medium forged not very long ago. The possibilities of fiction in the new medium were still largely untried. In Marathi there was an earlier tradition of prose writing by the Mahanubhav sect going back to the thirteenth and fourteenth centuries, but the language in which Marathi novels came to be written in the mid-nineteenth century had its beginning only in the same century. The first essays in novel writing in India entailed a two-fold adventure: experimenting to create a narrative form previously not part of the Indian literary heritage, and writing in a medium hitherto largely untested as a mode of literary expression. Compounding all these difficulties was the basic fact that the novel developed in India first in a colonial situation where the absolute superiority of everything published in English was taken for granted. It is perhaps unfortunate that the nineteenth-century Indian novelist had as his model primarily the British Victorian novel; with hindsight after a century it seems the British model was perhaps the least suitable for the Indian mind in the nineteenth century. The brooding inwardness and philosophical quality of the nineteenth-century Russian novel or the intensely moral preoccupation of the nineteenth-century American writer might have demonstrated to early practitioners of Indian fiction alternative modes of writing novels. A number of creative writers in our own time have remarked how little they have been influenced by English literature and how much by European and American literature, and of late by Latin American literature.[23]

India's first generation novelists had hardly any access to Tolstoy, Melville or Flaubert. With total servility they imitated

mediocre English novels, often devaluing their own talents in
the process. Mention has already been made of O. Chandu
Menon trying to adapt a novel (which has almost been
forgotten today) by Disraeli and ending up writing a genuine
first novel in Malayalam. Harinarain Apte, the first major
novelist of Marathi, thought it worth his while to translate a
rather trivial novel called *The Seamstress* by G.W.M. Reynolds.
Bengali critics foisted the epithet 'the Scott of Bengal' upon
Bankim as a supreme honour, while in actual fact Bankim as
a novelist, if not as a historical novelist, is more complex and
original than Scott. Realism came to be held as the highest mode
of perception (see Appendix II) and a good word from the
English press the highest conceivable reward. Pearychand
Mitra, author of a remarkable early novel in Bengali, cherished
ambitions of being published in England but was fortunately
dissuaded by his friend E.B. Cowell. Pearychand's son Chunilal
Mitra wrote a satiric sketch along the lines of his father's *Alaler
Gharer Dulal* and called it *Kolikatar Nukochuri* (*Hide and
Seek in Calcutta*) (1869), but also felt compelled to call it
Mysteries of Calcutta in English on the title page. This
consciously echoed Reynolds whose multi-volume *Mysteries of
London* was very popular reading in nineteenth-century India.

However, in spite of the limitations of the model, in spite of
basic incompatibilities between the English and the Indian
temperaments, the novel in India which began under English
tutelage soon began to acquire its own distinctive character.
Without attempting to arrive at any definition of the Indian
novel, it will be our purpose to examine the synthesis of a
borrowed literary form and indigenous aesthetic—as well as
cultural expectations—in order to determine the extent to which
the form has undergone mutation in the process.

II

PILGRIM PROSE AND THE NOVEL OF PURPOSE

While acknowledging that the development of the novel in India is an indirect result of the spread of English education and consequent exposure to Victorian literature, we tend to overlook the fact that the prose in which the early Indian novels came to be written was also shaped to some extent by European colonial enterprise. Not only was an appropriate prose medium brought into being, means were also devised to produce prose works in multiple copies for wider circulation.[1]

In this connection the establishment of the Serampore Mission Press in 1800, William Carey's association with it, and the founding of the Fort William College in the same year formed a conjunction of crucial importance. At Serampore Carey enlisted the help of Indian scholars—Ramram Basu for Bengali, Atmaram Sarma for Assamese, Vaijnath Sharma for Marathi, among many others—to translate the Bible into Indian languages. Sometime earlier, Bengali as well as Persian characters had been designed and cast into type by Charles Wilkins, an officer of the East India Company, and the first Bengali books had already been printed[2]—these being legal codes. Fort William College decided to teach 'Oriental' languages (which included Arabic and Persian) to the newly arrived officers of the Company, and Indian language writers were recruited for this purpose by the College. Thus, missionary enterprise in Indian languages was supported by administrative efforts.

As far as the evolution of literary prose was concerned the missionary efforts seemed at first to lead up a blind alley. It is well known that although the Bible was translated into many Indian languages by the early years of the nineteenth century,

none of the translations had much literary value. The magic of
the Authorized Version (1611) in English, which was also a
translation, was never repeated in an Indian language. The use
of prose for literary purposes was not very common at this time
in India, hence the translators had to forge a new medium as
they went along. Not being particularly creative or imaginative
(they were either enthusiastic missionaries or erudite scholars
in the local languages), the nuances and poetic possibilities of
the language eluded them. The stiff quaintness of Christian
prose would later become a proverbial butt of ridicule in most
Indian languages.

Secular prose got off to a better start, and some of the earliest
Bengali works of prose fiction—such as Ramram Basu's
Pratapaditya-charitra (1801) or Mrityunjoy Tarkalankar's
Prabodha chandrika (1833) — were written to provide the Fort
William College students with reading matter. Dr John Gilchrist
of the College collected Urdu and Hindi writers from different
parts of North India, and the best known Urdu prose narrative
written for text book use by the college is *Bagh-O-Bahar* by Mir
Amman (1801). This was written in a simple style, later called
'Fort William Urdu', which was looked down upon by the
literati in and around Delhi and Lucknow as a style too
direct and functional to be literary. The Hindi pundits that
Gilchrist brought to Calcutta came to be called *bhakha munshis*
because 'the word bhasa or bhaka (meaning language) had
been used loosely by the Muslims of northern India to denote
the various Hindi dialects.'³ Hindi was not yet a standardized
language; Brajbhasa and Avadhi were rich in poetic tradition,
but Khari-boli was what the bhakha munshis were advised to
develop. We are told that Gilchrist had a preference for the use
of Arabic and Persian vocabulary in Hindi. But when Capt.
William Price took charge of this department of the College in
1824 he emphasized Hindi rather than 'Hindustani' (Hindi
mingled with Urdu).⁴ Writers like Lalloo Lal (1763–1835) and
Sadal Mishra (1768–1848) produced translations as well as
some original works. Lalloo Lal's *Premsagar* (first publication
probably in 1802) and Sadal Mishra's *Nasiketopakhyan* (1803)
are among the earliest publications in Hindi prose. Only the

celebrated *Rani Ketaki ki Kahani* (1801) by Insha Allah Khan was written outside the Fort William orbit and missionary ambience.

The Bible apart, translating *Pilgrim's Progress* appears to have been one of the commonest and most popular missionary activities in every part of India. The Tamil translation appeared as early as 1793, the Kannada rendering in 1841, the Malayalam in 1845; the Assamese translation, entitled *Yatrikarar Yatra*, was serialized in the Baptist Mission periodical *Arunodoy* which started in the 1830s. The Marathi version by Hari Keshav, entitled *Yatrik Kraman*, is said to have inspired Baba Padmanji's *Yamuna Paryatan* (1857), an early novel in Marathi.

Thus it is possible to regard the fiction written by Christians—whether originally so or converts—as a recognizable product of mid-nineteenth-century India. Among the early crop were *Phulmoni-O-Karunar Bibaran* by Hannah Catherine Mullens in Bengali (1852), the earlier mentioned *Yamuna Paryatan* in Marathi, and *The Slayer Slain* by Mrs Collins in English (1864–6; translated into Malayalam in 1877). The first two are sometimes hailed as the very first novels of these languages; the third when translated achieved this status for Malayalam. Similar claims have been made for *Kamini Kanta* (1877; Assamese) written by G. S. Gurney, a Christian missionary, and for *Prathapa Mudaliar Charitram* (1879; Tamil) by Samuel Vedanayakam Pillai, a third generation Christian. If not the first novel, *Sukumari* (1897), an early Malayalam novel by Joseph Muliyil, seeks to demonstrate the benefits available to Indians by converting to Christianity.

The whole body of such works constitutes Christianity's contribution to the development of the novel in India. A comparative discussion of the themes and their presentation in some of these novels will enable us to assess the nature of this contribution.

II

Hannah Catherine Mullens, who lived and died in Calcutta (1824–67), and Mrs Collins of Kottayam (d. 1862) were contemporaries whose lives ran roughly parallel courses more than a thousand miles apart, although they could not have been aware of each others' existence. They were daughters and wives of well-known English missionaries[5] and devoted their lives to helping their husbands in the work of proselytizing and educating Indians, as Jane Eyre would have done had she married St John Rivers. Mrs Mullen's brief biography attached to her book reads like a page out of a Victorian novel: the devoted daughter and dutiful wife whose endless task of child-bearing and social obligation in a hostile climate is punctuated by the secret pleasures of reading and writing—secret because her stern father thought writing was an indulgence and because she herself thought she derived too much pleasure from these activities. One is reminded of Dorothea Brooke in George Eliot's *Middlemarch*, a fictional contemporary of Mrs Mullens, who 'enjoyed riding so much in a pagan and sensuous way' that she always looked forward to renouncing it. Not too much of this pleasure, however, is evident in the utilitarian tone of her book—*Phulmoni-O-Karunar Bibaran*—which was written 'for the education of the native Christian women'.

Mrs Collins' book *The Slayer Slain* on the other hand shows a deeper awareness of the physical reality in which her moral tale is set. The landscape and atmosphere of Kerala is evoked in concrete detail—the splash of an oar as a visitor approaches the house, the swish of wind through paddy fields, the shade of the mango tree under which children sit on reed mats and squeeze juice out of ripe mangoes, the yard of the landlord's house lush with tendrils of yam climbing jack trees, yellow flowers of the pumpkin and cucumber, 'deep purple leaves of the *cheeras* standing in contrast to the green ridge of *goorkas*'. Mrs Collins' novel is full of the kind of particularization and naming of specific objects that is seldom to be found in the early fiction written by Indians for similar instructive purposes. She was employing a western realistic mode in her handling of Indian material, whereas in traditional narrative fiction descriptions of landscape or nature tend to appear in a stylized manner, more to satisfy literary convention than as the actual observation of specific detail. Mrs Collins is so preoccupied with the flora of tropical India that she occasionally digresses from her main narrative to comment on wild flowers never found in gardens, and exhorts her reader—

while we gaze and admire the festoons of the climbing yam, or the massive vegetation that overtops the chana, one cannot but regret that the gay and lovely flowers which adorn and beautify the homes of the bulky elephant and creep round the den of the fierce tiger find no fostering hand of man. The lovely purple blossoms of the Kakapoo or the large pure white of the *Munda walli* seldom meet with a gentle hand to transplant them from the wilds of the jungle. Oh! why not, ye mothers and daughters of India, transplant some of these gems of nature to your cottage doors...[6]

Catherine Hannah Mullens' apprehension of the landscape of Bengal is never very vivid in *Phulmoni-O-Karuna,* though she pays detailed attention to human habitations and interiors of houses. Phulmoni's yard is freshly swept and a contented cow and calf stand in a shed which is covered with a creeper heavy with green gourds. Through the kitchen door one can glimpse the gleam of well-scrubbed vessels. Of the 'ten or twelve plants' in her courtyard, some are herbal, the rest are meticulously named—tulsi, gandharaj, etc.—and there is finally a touch of luxury: a China rose creeper, planted by Phulmoni's daughter, which serves as a recurring motif throughout the narrative.

These are minor details, but taken as a whole they mark a departure from the narrative tradition of Indian culture in which such specificity in the recording of a physical environment had never been thought important. Though second generation residents of India, these English women were inheritors of a different literary tradition—one in which individual perception of an object is more important than its essence, the particular more valid than the universal. The books written by Mrs Mullens and Mrs Collins may not be worth serious literary attention but they help to illustrate how realism as a literary technique is connected with a particular world-view. Some of Ian Watt's remarks from a different context may be recalled here to underline this point:

just as there is a basic congruity between the non-realist nature of the literary forms of the Greeks . . . and their philosophical preference

for the universal, so the modern novel is closely allied on the one hand
to the realist epistemology of the modern period, and on the other to
the individualism of its social structure. In the literary, the philosophical
and the social spheres alike the classical focus on the ideal, the universal and the corporate has shifted completely, and the modern field
of vision is mainly occupied by the discrete particular, the directly
apprehended sensum, and the autonomous individual.[7]

Though simplistic in their intention and execution, the works
by these English women prefigure in a rudimentary way the
use of realism in the fictional handling of Indian reality. These
books are not mere exempla exhorting the natives to become
Christians. In fact neither of them is primarily concerned with
conversion. In *The Slayer Slain* the conflict is not between
Hinduism and Christianity, but between Syrianism, an early
form of Christianity in Kerala, and the Protestant Church. Even
those already converted to Protestantism tended to slip back to
the old religion, and the apostasy of Koshy Kurien, a rich
landlord, is the central concern of the novel. In *Phulmoni-O-
Karuna* most of the characters are already Christians and are
being made aware of 'the practical influence of Christianity on
the various details of domestic life'. Two conversions are
referred to during the course of the story, but they are incidental
to the main plot. As indicated by the title, the plot revolves
round two women and their families in a 'mofussil town in
Bengal by a river'. Phulmoni, though poor in means, is rich in
faith and leads an exemplary life. In contrast, Karuna is lazy and
indifferent to church-going. Such contrasts are familiar in
folk and fairy tale traditions: it is interesting that Nazir Ahmad's
Urdu novel *Mirat-ul-Arus* (1869) and Pandit Gauri Dutt's
Hindi narrative *Devrani Jethani ki Kahani* (1870), which are
also stories meant chiefly to instruct women how to conduct
themselves, depend on similar devices of contrast. In Nazir
Ahmad's novel Akbari is lazy, querulous and selfish and her
sister Asghari is intelligent and efficient. In Gauri Dutt's story of a
bania family in Meerut, the elder daughter-in-law is illiterate and
jealous while the educated and responsible younger daughter-
in-law brings happiness to the family. Behind all these didactic
tales one can detect dual archetypes of numerous folk
tales

In Bengali, for example, there is the familiar story of Sukhu and Dukhu, the two sisters, who made the same journey and were given the same boons, but one ended up beautiful, happy and rich and the other ugly, poor and miserable because the latter was greedy.

Phulmoni's house is neat, her children well-scrubbed and polite. Karuna leads a disorderly life; she loves to gossip; her husband beats her regularly after coming home drunk; her children tell lies and steal. Interestingly as a character it is Karuna who becomes more alive in her crude and aggravating ways than Phulmoni the paragon of virtue. The author uses the persona of the District Magistrate's wife to narrate the story. Her acquaintance with Phulmoni and Karuna and her gradual involvement with their families is the main story line. The relationships grow; through births and deaths and festivals and disasters the narrator gets to know them intimately. Peripheral characters are introduced and the story ends happily with one marriage, one conversion and Karuna's transformation through suffering into a true Christian.

The Slayer Slain is located 'in the province of Travancore in the town of U—', where Koshy Kurien torments his Christian serfs because they refuse to work on the sabbath day. In an angry exchange he accidentally kills the little grandson of old Paulusa, a bonded labourer in his farm. Thereafter the story is about his pangs of conscience which he hides under further violence, and his young daughter's attempts to change his heart. A few sub-plots are introduced, and all converge in the happy ending.

The two novels have a few obvious similarities—the endings, for example. In both novels a young girl (Mariam in *The Slayer Slain* is fourteen and Sundari in the Bengali novel is fifteen) provides the Christian norm and becomes the agent of regeneration. In both we find marriages willed by the young people themselves rather than by their parents, perhaps the earliest fictional representation of such individualistic marital choice exercised in modern India. And in both novels children play an important part in disseminating education and Christian ideals.

But in spite of being written by women of such similar backgrounds and being almost contemporaneous in time, the two

novels have significant differences. The point of view in
Phulmoni-O-Karuna is that of a benevolent white woman who
can act as *deus ex machina* by offering jobs or money to the
poor villagers in their moments of crisis. These men and
women belong to an isolated community where to become the
domestic servants of English families is the height of ambition.
These people are seen as unrelated to the rest of Bengali society,
although the narrator does lecture Phulmoni on the need to
remain a Bengali and not to ape the ways of the British.

No Phulmoni . . . I do not want Bengali women to behave like English
ladies. When they speak to men they do require a certain kind of
modesty—which can be achieved by making the heart pure rather than
by drawing a veil over the face.

She also recommends Bengali names to be given to children
rather than 'English names which the natives cannot generally
pronounce'.[8] The author even provides a list of Bengali names
which have no reference to 'the idolatrous objects of Hindu
worship.'[9] *Phulmoni-O-Karuna* thus also serves as a manual
containing a list of names, a summary of the gospel plan of
salvation, practical advice about what to do at childbirth and
sickness, how to keep the sabbath and repudiate superstition.

The didactic purpose does not, however, necessarily obscure
the human aspect. Karuna is loud, lazy and untidy. If she has
one paisa she will spend it on tobacco rather than on soap, yet
she is real as a character. She forgets all her resolutions
to be gentle when her husband comes home drunk. One day
when she does force herself to be tender and submissive, the
drunken husband is so taken aback that he decides he must be in
a brothel. He mumbles as he falls asleep: 'This one has a kind
voice, I think I'll visit her more often.' These are places where
the novel, transcending its didactic purpose, becomes human
and even comic.

Yet Mrs Mullens' novel seems limited and narrow when
compared with Mrs Collins' *The Slayer Slain* which deals with
people who are connected with land, for this association gives
their lives more vitality. There is something abiding and
elemental in this story of guilt and retribution. The remorse of
Koshy Kurien, however melodramatically presented, has an

element of real conflict. The slave Paulusa whose grandchild he has killed saved Kurien's daughter from drowning. Before disappearing under the water, Paulusa the slave says'You killed my child, I have saved yours. We are equal now'.

The last words of the slave haunted him wherever he went. He saw them written in large characters on the waters of the deep flowing river. He saw them on the paddy banks and heard them in the humming sound of the water wheel. The winds seemed to carry the words on their wings and creep through the crevices of the door as he tossed about on his sleepless cot. And the fearful crime of a double murder hung over him like an unsheathed sword, and drove from his breast happiness and peace.

For a good part of the novel Koshy Kurien is haunted by his guilt, but when the resolution comes it is too facile to be psychologically convincing.

Are these Christian novels historical curiosities for us today, or are they part of our literary tradition? For *The Slayer Slain* a literary historian, Krishna Chaitanya, has claimed that 'it features symbolic anticipations of the directions in which the currents of relatively greater vitality in Malayalam fiction would flow later.'[10] He also points out that this is perhaps the first novel in India to deal with the exploitation of landless labourers, a concern that recurs in Indian fiction of the twentieth century.

III

What has been identified as the 'Christian' novel was also generally an early work of fiction in the language concerned and is often cited as the first novel written in that language. This claim has been made on behalf of *Phulmoni-O-Karuna* by Chittaranjan Banerjee[11] on three grounds: it has original subject matter, i.e. it does not retell an old myth or puranic story; it creates a detailed and realistic setting; it portrays authentic characters. But in the opinion of Saroj Bandopadhyay, one of the most perceptive fiction critics in Bengali, *Phulmoni-O-Karuna* does not fulfil any *one* of the three primary conditions of a novel: namely a vision of life, a pattern of experience against a wide cultural and historical perspective, and a situation of conflict.[12] It is the second of these conditions which is probably decisive, and the Marathi narrative *Yamuna Paryatan* (1857) meets it a shade better than *Phulmoni-O-Karuna*.

Baba Padmanji's Yamuna is an unusual young girl in that she has been to school (run by Christians), knows how to read and write, and feels stifled by the values of orthodox Hindu society. She is married to Vinayak, a young man sympathetic to her ideas, but the family to which he belongs shares the hypocritical and rigid values of the rest of the brahman community. Yamuna's honesty and innocence create problems in this set up. When the servant is hungry and asks for food Yamuna's mother-in-law wants her to tell him that there is no left-over food. But Yamuna who cannot tell a lie says: 'there is left-over food but I am not supposed to give it to you'. Yamuna is horrified when a man in the neighbourhood dies and his wife's head has to be shaved because it was supposedly her bad luck which brought death to her husband. The woman commits suicide rather than go through this ignominy and Yamuna is afraid that this will be her fate if Vinayak dies.

Thereafter the book is a string of episodes of which Yamuna and Vinayak are a part during their journey across Maharashtra, and almost every incident touches upon the predicament of a widow. In Nagpur, Pandharpur, Nasik—wherever they go—they meet some unfortunate widow leading a miserable existence. There is Venu who is treated like a servant in a joint family, Daulat's widowed cousin who is made pregnant by a sadhu in the name of religious practice, and another widow who drowned her illegitimate child in the river. Running through all this there is a discourse about the desirability of widow remarriage. An indignant brahman widow tells Yamuna that remarriage is not for brahmans though it might do for other castes. Vinayak takes part in a formal debate on the subject where pundits quote scriptures for and against the remarriage of widows.

This seeming preoccupation with widow remarriage not only leads up to Yamuna's own later predicament; it also has topical relevance in view of the heat being generated over this particular issue in Bengal as well as Maharashtra at that time.

One of its main crusaders in Bengal, Ishwarchandra Vidyasagar (1820–91), finally got a bill passed in 1856—one year before the publication of *Yamuna Paryatan*—which made the remarriage of Hindu widows lawful in British-administered parts of the country. Enactment of the law did not, however, put an end to the controversy. We find in *Vishavriksa* (1873) Bankim scoffing at a character who had

become a bit of a village oracle. He had read *The Citizen of the World* as well as *The Spectator,* and it was rumoured that he had got through the first three books of geometry... He used to go round telling people, 'Give up worshipping brick and rubble. Get your old aunts married again. Teach your women to read and write. Let them out of their cages.' A special reason of his liberal attitude towards women was that there were none in his own household.[13]

Even as late as the 1930s a popular novelist like Saratchandra Chatterjee did not dare show a remarried widow in any of his novels. But the problem of Hindu widows had been very much the concern of writers in India from the mid-nineteenth century. Even in a novel as rudimentary as Pandit Gauri Dutt's *Devrani Jethani ki Kahani* (Hindi; 1870) there is reference to a nine-year-old girl whose husband died by falling off the terrace while flying kites; the good 'devrani' feels very sorry for the luckless girl:

The girls who play with her eat well and wear nice clothes. They laugh and sing. Doesn't she also want to be like them? Why should the burden of the seven rounds always weigh her down? It is the problem of our community only. Among the Muslims and among the British remarriage is possible. Now even the Bengalis are doing it. There is no bar to remarriage among the jats, gujars, barbers, kahars, aheers and dhobis...[14]

In *Yamuna Paryatan* an entire rambling chapter called 'Sabha' and a subsequent chapter called 'Nirnay' are devoted to the technical aspects of this issue, dwelling in detail upon what the Dharmashastras have to say. Vinayak takes the liberal standpoint in this debate.

Rather abruptly in the story Vinayak is killed in a bullock-cart accident. All the tales of the torture of widows repeated throughout the book now become real as Yamuna herself becomes a widow. She is ill-treated by Vinayak's family but is helped by her friend from Pandharpur and the friend's son Shivaram. Shivaram becomes convinced of the superiority of Christian values and seeks conversion. Yamuna too becomes a Christian and marries again. The novel ends when Shivaram starts an organization called 'Punarvivah Uttejak Mandali' (Association for the Encouragement of Remarriage) and 'gets help from England, Scotland and America'.

As a narrative *Yamuna Paryatan* is more perfunctory than the stories by Mrs Mullens or Mrs Collins and it does not even have the vividness of realistic detail that enliven those two works. There is no attempt at characterization nor a desire to integrate the separate episodes into a larger pattern. Yamuna and Vinayak remain mere recorders of events rather than actual human beings, and their picaresque journey all across Maharashtra has no ostensible reason except as a pretext to witness the plight of women. It is obvious that Baba Padmanji's interest was not in the creation of a new genre, but in promoting social awareness. He was already the author of some fifty pamphlets on various subjects of topical interest and writing a long narrative was for him just another method of influencing public opinion. Christianity is not the central issue in *Yamuna Paryatan*. It is offered as a practical solution to Yamuna's misery only after all other possibilities are explored in scholastic debates. Similarly, Samuel Vedanayakam Pillai, author of the first Tamil novel, had earlier written non-narrative verse and used the narrative form in his prose work for propagating the same moral ideas.[15]

The Malayalam novel *Sukumari* (1894) by Joseph Mulivil deals with conversion to Christianity in a more central way. The characters in this novel belong to a Thiya family who embrace Christianity one by one and under different circumstances. A girl called Chirutha, whose mother is dead and whose father had joined the Christian mission, is converted to Christianity at the death-bed request of an aunt, and Chirutha's name is changed to Sukumari. She is brought up in the Basel Mission and is helped by, among others, her father, now called Sadheart. The novel is important not for its literary value but for its reformist zeal and its insistence on education as a beneficial force and Christianity as a cure for social evils. It has a historical

value in that it gives a clear picture of the activities of the Basel
Mission from 1841, the year in which they established a branch
at Cannanore (the place where Dr Gundut, the grandfather
of Herman Hesse, once worked).

These Christian novels in different Indian languages have
hardly ever been linked, perhaps because the impact of
Christianity on Indian literature has generally been regarded as
marginal. Compared to several other British colonies the direct
influence of Christianity on Indian society has indeed been
insignificant. Chinua Achebe's novels *Arrow of God* (1964) or
Things Fall Apart (1958) are moving documents of the
confrontation of two religions, two ways of life in Nigeria, as
are Ngugi wa Thiongo's two early novels accounts of similar
encounters in Kenya. No records like these are available in
Indian literature. Although Western ideas transformed the
intellectual climate in India, the Christian religious experience
did not affect the mainstream culture in a big way. There has
been some remarkable non-fiction prose in Marathi, specially
Baba Padmanji's autobiography *Arunoday* where he talks of
his conversion to Christianity in a serious and self-searching
manner. Lakshmibai Tilak's *Smruti Chitra* (translated from
Marathi into English as *I Follow After*) also deals with the
experience of Christianity in an intense and personal way. But
by and large in the imaginative literature of modern India, the
spiritual and emotional aspects of Christianity do not seem to
have found any expression. Bengal's major nineteenth-century
poet Michael Madhusudan Dutt consciously sought and
obtained conversion to Christianity, but his poetry bears no trace
of any inner tension his conversion might have caused, nor has
he left behind any prose recording his faith. Of the two
missionary women discussed earlier, Mrs Mullens regarded
Christianity merely as a practical guide to conduct while Mrs
Collins, although she did touch upon the theme of guilt and
regeneration, did not go deep into it. Baba Padmanji in *Yamuna
Paryatan* dealt with Christianity only because it offered an
escape from the tyranny of an oppressive society. This aspect of
Christianity has continued to be relevant in Indian life and
literature. In one of the novels of Saratchandra in the next
century (*Srikanta*, Part III, 1927) there is passing reference to a

family that has been degraded because of a lapse on the part of an ancestor. Harassed and oppressed by society, a woman in that family says: 'I have an uncle who has become a Christian after getting a job in Dumka. He has escaped this torture.' But such drastic solutions are never considered seriously in Saratchandra's novels. In the well-known Kannada novel by B. Shivaram Karanth, *Chommana Doodi* (1945), conversion to Christianity is offered as an attractive alternative to social oppression, but the central character Choma prefers to retain his lower caste Hindu identity although his children succumb to the temptation.

Of the two kinds of people who were converted to Christianity in nineteenth-century India, intellectuals like Michael Madhusudan Dutt or Baba Padmanji were in a minority. The majority were from the socially and economically depressed classes who found in Christianity a better way of life. The situation regarding conversion is expressed by an old grandmother in *The Slayer Slain*—'whoever has heard of a Brahmin getting converted to Christianity?'—and Krishna Chaitanya has pointed out an interesting paradox in this professedly Christian novel: 'While it attacks caste, the social ambience has insidiously influenced the author with the prestige of caste; that is why the Brahmin comes on the scene to donate a pedigree to the (Christian) heroine by claiming a blood relationship.'[16]

The task of the Christian missionary in India was an uphill one in the early years of the nineteenth century. The Abbe Dubois, after many years of missionary work in South India, admitted in 1815 that he had 'watered the soil with his tears... They had "fallen on naked rock"... In the last thirty years there had been only 300 converts of whom 200 were Pariahs'.[17] Seen as a whole, in the nineteenth century the spread of Christianity in India was not a central event socially or culturally. The literary by-products of missionary activities also turn out to be minor works, contributing only indirectly to the evolution of the novel in India, but the tributaries are important in understanding the mainstream. The historical value of these works cannot be entirely ignored because they introduced a non-traditional element in the narrative and brought in the first traces of realism, even though their primary aim was didactic.

In a recent essay, the contemporary Marathi novelist Bhalchandra Nemade has divided the entire corpus of Marathi fiction into three basic strands that have evolved, diverged, combined and parted to form many patterns.[18] One of them he calls the *Yamuna Paryatan* strand, which incorporates social reform impulses in a functional manner in literature. The other two are the *Muktamala* (1861) and the *Mochangad* (1872) strands, wherein these two early novels embody the imaginative-romantic urge and the revivalist-historical spirit respectively. Similar trifurcations can probably be made in the fiction corpus of other Indian languages. What we may note here is that the first of these, manifested in utilitarian or didactic aims in some of India's early fiction, was not confined to Christian writers. The early novels of Nazir Ahmad (1836–1912), for example, were written mainly to instruct. Ahmad's first work, *Mirat-ul-Arus* (*The Bride's Mirror*; Urdu, 1869), was written to show young girls the qualities that would stand them in good stead when they set up house. One year later Pandit Gauri Dutt's Hindi narrative *Devrani Jethani ki Kahani,* sometimes claimed as the first novel in Hindi, had an identical theme, although the differences in social and cultural milieu and in language—one is about a Hindu bania family in Meerut and the other about a Muslim family in Delhi; one is written in the dialect of Khari boli which was actually spoken in western UP at that time, rendered without any punctuation or paragraphing, and the other in a flexible near-colloquial Urdu, graceful yet functional, the language of the Muslims in Delhi—make the total effects somewhat dissimilar. But both stories describe two brides: one lazy and worthless and the other educated and exemplary—and the second girl in each case sets up an informal school for the dissemination of education among women. Since most of these instructive books were published, if not written, with official patronage (*Mirat-ul-Arus,* for example, was originally written for private circulation), they were often prescribed as textbooks. Nazir Ahmad's *Taubut-un-Nasuh* (1877), which was translated by M. Kempson into English (*The Repentance of Nusooh,* 1884), carried a foreword which ends thus:

for all these reasons I have great satisfaction in commending Mr Kempson's translation to all who are interested in India, and also in advocating...the use of the original treatise as a text book for the acquisition of Hindusthani and for the examination of proficiency in the same.   W. M., July 1884.

This Kempson was obviously the Director of Public Instruction at the time because Gauri Dutt, the author of *Devrani Jethani ki Kahani,* joyfully informs the reader in his preface how Mr Kempson, finding the book very enjoyable 'had ordered it to be published with corrections and bought two hundred copies of it'. *Mirat-ul-Arus* was written by Ahmad for his daughters and the women of the *mohulla* also 'came to listen when it was being read'. When his eldest daughter got married Ahmad 'included the manuscript in her dower as a jewel of great price and it achieved no less a reputation in her new home'.[19] The book was published only after its chance discovery by Kempson, in whose department Ahmad was then employed.

Both the works of Gauri Dutt and Nazir Ahmad emphasize the need for education among women to make family life happy: 'When the man is literate and the woman is not, there can be no meeting of minds,' wrote Dutt, and thus Anandi, the younger brother's wife, is shown to bring cheer and prosperity into the family. In Ahmad's story Asghari, the younger sister, has a similar effect on a family which was nearly destroyed by her foolish and ignorant older sister Akbari. Asghari is very articulate about dignity of labour (p. 150), argues against pomp at weddings (pp. 165–7) and has intelligent views about British rule in India and employment opportunities (p. 138). She has the courage and initiative to travel alone from Delhi to Sialkot in order to bring an errant husband back to the right path. In the informal school she ran in her house, girls were taught to read, keep accounts, do calligraphy, stitch, dye clothes, embroider (twelve different kinds of fancy needlework are mentioned), cook (she mentions eleven varieties of pulao); child care and home remedies of common ailments are also listed, and recipes and remedies are woven into the texture of the story so that the reader can learn while being entertained. Similar devices are used in *Devrani Jethani ki Kahani,* and in both stories several letters are reproduced, perhaps with the intention of making women readers learn how to write letters properly. Special mention must be made of the letters these girls received from their fathers in both novels, advising them how to conduct themselves in their new homes. These letters distil the social values of the milieu with emphasis on a woman's tolerance, self-effacement, co-operation and consideration for others. We have here entertainment, combined with practical education as in Mrs Mullens' novel, and the same pattern continues in all the didactic literature of the period. At least one Englishman saw the influence of Christianity behind this kind of writing. In the foreword to Ahmad's *Taubut-un-Nasub*, which was written to counter indifference to Islamic religion, he self-righteously claimed that even though the book propagated the practice of Islam it was actually written under Christian influence: 'In fact it is only in country [*sic*] under Christian influences, like those which happily are seen and felt in India, that the idea of such a book should present itself to the Moslem mind.'[20]

Although the aim of *Taubut-un-Nasub* is to emphasize the need for religious practice, the book is more than a mere didactic tract. The novel begins with an evocative description of a cholera epidemic (incidentally, epidemics were very much a part of ordinary reality; the first page of Gauri Dutt's story casually mentions 'the last cholera season' when both parents of Sarvasukh died) which nearly killed Nasuh. Nasuh had a vision in which his dead father told him how he repented after having led an irreligious life. Nasuh becomes a devout Muslim when he recovers, and the rest of the narrative describes his attempts to convince his wife and children of the need to be good Muslims. He meets resistance from only two of his five children, specially from his eldest son who is a poet, a kite-flier, a pigeon-keeper and a reader of secular books like *Gul Bakavali*, *Fisana Ajaib* and *Chahar Darvish*—a rather vivid portrait of a dilettante. A long extract from the first chapter will illustrate how the narrative operates on two levels, the evocative and the pragmatic:

Not very long ago the cholera was so bad in Delhi that thirty or forty deaths a day were counted in a single street. The public thoroughfares, once crowded with men till midnight, were empty. The hammer of the artizan had ceased, and the street criers were dumb. There was no visiting, no hospitality, no friendly intercourse, for all had lost heart and hope. A man might be walking about in the best of health, when all of a sudden he felt ill. There was no chance of escape, no time even to make a will. In one short quarter of an hour all was over—vomitting, medicine, prayer and the death rattle. The city lost half its population.

Nusooh, whose story I am about to tell, had forseen at the first outbreak the necessity for precaution. He had his cooking vessels re-tinned, and impressed the duty of cleanliness on his household. Frankincense was burned, and camphor and charcoal placed about the rooms. The usual native remedies—coconut, aniseed, tamarind, lime juice, etc.—were provided in readiness for an emergency. A stock of English medicines was laid in—cholera-pills from the dispensary, tincture from Allahabad, chlorodyne from Agra, and a specific, said to have been recently discovered by a Bengali, was procured from Benares...[21]

The first paragraph is a brief but effective evocation of pestilence, bringing to mind Chaucer's *Pardoner's Tale*. The second has a manual-like quality, enumerating specific remedies and actual places in a pragmatic matter-of-fact manner. The two qualities—the once-upon-a-time aura and the here-and-now flavour—alternate in the entire text of *Taubut-un-Nusuh*.

Devrani Jethani ki Kahani has very little evocation, the intention of the writer being merely 'to show the customs of the banias at the time of birth, death and marriages; what are the differences between educated and illiterate women; how to bring up children...etc.' (Author's Preface to the first edition). The other contender for the honour of being the first Hindi novel, *Pariksha guru* (1882) by Srinivas Das also has a didactic purpose: largely the protection of Hindu values against the influence of the West. Nazir Ahmad was an almost exact contemporary of Bankim, the first major Indian novelist. In Bankim one finds very little of the single track utilitarian purpose that characterizes much of the early Hindi/Urdu fiction. If in Bankim's later novels (for example, *Anandamath* or *Debi Chaudhurani*—discussed in the next chapter) Hindu values or the Shakti cult are valorised, this is not done at the cost of characterization or believable human situations.

Using fiction primarily as propaganda—be it for Christian, Muslim or Hindu values—has its own built-in problems. These efforts, however effective they might have been in their immediate non-literary purpose, could not contribute to the evolution of the novel in any central way because a subtle and certain link exists between the novel as a genre and liberal ideology as a way of life. This is because the novel recognizes the uniqueness of individuals and accepts a plurality of beliefs and values. In other words the novel comes into being only when the reader and the writer believe in the uniqueness of the individual and his will. Individuals and their relationships are important only when they are not completely controlled by and are not totally subservient to powers like the church or the Dharmashastra or the gods on Olympus.

'The first great novel of world literature stands at the beginning of the time when the Christian God began to forsake the world', said Lukacs in a slightly different context,[22] but one can see that individual doubt and the loneliness of man are conditions more conducive to the writing of the novel than absolute faith or total dependence on collective values. This may be one reason why even in our own time very few 'committed' writers have been able to create memorable works of fiction. A character who has not made the final choice yet, either in terms of spiritual faith or of political creed, can provide the writer with richer fictional material than one who has already identified himself with a set of beliefs. In a recent interview[23] Satyajit Ray corroborated this view in terms of the cinema while explaining why in his film *Pratidwandi* (*The Adversary*) the Naxalite brother is less important to him as a psychological entity than the vacillating older brother torn by doubts. In real life firm commitment may lead to prompt and decisive action, but rendered in art this tends to reduce human figures to puppets. The novel as a genre needs especially to examine human beings in situations where individuals exercise a choice. The purely missionary or propaganda novel which does not allow such choice is a contradiction in terms. The early Indian novels of the variety discussed here engage our attention largely for historical reasons, as examples of the earliest attempts to graft a new form on existing fictional traditions. Each such attempt was also an exercise in narrative prose which extended the possibilities of generic development.

III
RECREATING A PAST:
FICTION AND FANTASY

The earnest novels of purpose discussed in the previous chapter were never as popular or numerous as the lavish novels about heroism and adventure, love and romance, that captured popular imagination by the end of the nineteenth century. Elements of fantasy and intimations of history are inextricably tangled in these works. Chronicles merge with legend, events lapse into magical happenings, and kings who lived once-upon-a-time cast their spell upon those who ruled at a specific period and over a definite area. The spectrum covered by these novels is not only wide but also colourful and varied. It ranges from Romesh Chandra Dutt's *Maharashtra Jiban Prabhat* (Bengali; 1878)—a fictional rendering of the rise of Maratha power under Shivaji underlaid by the story of a romance between a *killedar's* daughter and a young *havildar* on the one hand, and on the other Devakinandan Khatri's *Chandrakanta* (Hindi; 1891), a fabulous saga of the sensational and the marvellous. In between these poles of fiction and fantasy were novels like Bankimchandra Chatterji's *Rajsingha* (Bengali; 1882) in which some of the major characters—such as the emperor Aurangzeb, his daughter Zebunnisa, Rajsingha the Rana of Udaipur—are drawn from the pages of history, but their human dimensions are depicted through their relationship with non-historical figures; Ramchandra Bhikaji Gunjikar's *Mochangarh* (Marathi; 1871), where the setting is a hill fort in Maharashtra and Shivaji's capture of the fort brings about a happy ending although the actual story revolves round imaginary characters; C.V. Raman Pillai's *Marthanda Varma* (Malayalam; 1891), where the feud over the throne of Vanad (Travancore) is the historical backdrop but the central figure of the romantic plot turns out to be an imaginary lieutenant of the prince; Umeshchandra Sarkar's *Padmavati* (Oriya; 1881), in which the accurately presented historical strand is overshadowed by the complicated love story, or Kishorilal Goswami's *Hriday Harini*, subtitled 'Adarsh Ramani' (Hindi; 1890) which, in spite of the fact that Lord Clive, Sirajuddoula, Mir Jafar and Amichand appear in person, is palpably a romance, weaving the supernatural and the fantastic in and out of its narrative strand.

Underlying the use and abuse of history in Indian fiction of this period there must be intimations of the cultural crisis that was overtaking Indian society as a result of the impact of Europe. One aspect of it had to do with coming to terms with the Indian past, recent as well as remote, and learning how to handle it. Mircea Eliade has suggested a correlation in all traditional societies between the secularization of culture and the emergence of historical consciousness.[1] According to him, mythic time of a pre-modern culture is cyclic, in which the same primordial drama is continually re-enacted; whereas historical time of modern man is linear, tracing in an irrevocable progression the events from the past passing on to the present and future. In literary terms, mythic time is represented by the *puranas*, perhaps also in the *akhyana* works composed in Sanskrit during historical periods; but historic time had yet to be represented in Indian writing. Rabindranath Tagore posed the same question a little differently in his statement that when a group of people are unified as a religious community they feel the need to record their faith—which is timeless; but only when they have a secular human bond does the desire to record their daily joys and sorrows and temporal events assume importance.[2] It is the latter desire which provides the staple of fiction, and certainly English education in India, at least in the urban centres, was creating an awareness of history and stressing the importance of temporal events as well as human bonds irrespective of faith.

Quite noticeably in the late nineteenth century, a new and unprecedented interest in history can be seen to be shared

between Indian writers of fiction and their readers. Several
interrelated factors may have contributed to this interest: (a)
general exposure to other cultures through the study of English;
(b) awareness that Indians were different from the British and
a consequent curiosity to understand the past that would account
for the present; (c) rediscovery of Indian history in books that
the British wrote about India; (d) the desire to rewrite these
accounts from an indigenous point of view. We know from
Pandit Haraprasad Shastri's account that Bankim was far more
interested in the study of history than of literature and that he
read avidly about Europe: 'He wished that there might be a
regeneration of life in Bengal along the lines of the European
Renaissance. He hoped one day to write a history of Bengal.'[3]
His fictional and non-fictional work is liberally strewn with
references to European history, especially of Italy. As for
Bankim's younger contemporary Romesh Chandra Dutt
(1848–1909), he once wrote: 'I do not know if Sir Walter Scott
gave me a taste for history or if my taste for history made me an
admirer of Scott, but no subject, not even poetry, had such a
hold upon me as history.'[4] We also have evidence that Harinarain
Apte (1864–1919), who wrote eight historical novels in Marathi,
was well-known for his copious reading of history, particularly
about the Maurya Empire, the Vijayanagar Kingdom (he read
Sewell's *A Forgotten Empire*) and, of course, the recent past of
Maharashtra (Grant Duff's history left him unsatisfied because
of certain distortions in the account of Shivaji's exploits).[5] This
newly-awakened interest in the past could not have been
unrelated to a nascent nationalism among the reading public at
large which the novelist could exploit. History could, if
necessary, be altered when it suited his purpose better. Thus
the writing of fiction based on history, with varying degrees of
historical accuracy, was a necessary part of the new situation.

Such writing was complicated by the existence of a pre-novel
narrative tradition which also depended, as did historical
fiction, on a setting remote in time, necessary—and quite
opposite in need to that of historical fiction—for a suspension
of disbelief. Popular reading in Hindi in the nineteenth century,
before the novel came into existence, included stories from

Betal-pachisi, Simhasan-battisi, Tilism-i-Hoshruba Daastan-e-Ameer Hamza, Kissa Tota-Maina and other tales featuring miraculous happenings and magical deeds.[6] In Bengali, too, popular fiction of the pre-Bankim era included material like *Gul-e-Bakavali* and *Bijoy Basanta,* as testified by Rabindranath in his essay on Bankim where the latter is credited with bringing Bengali literature out of such childish entertainment towards maturer pleasures.[7] The same *Gul-e-Bakavali* which Rabindranath read as a child was available to the Marathi reader in Navalkar's translation from Gujarati, the Gujarati version itself being a translation from Persian. A similar text, *Gul-va-Sanovat,* was translated by David Higham Divekar, a 'Bene Israel', in 1867 from Urdu into Marathi.[8] Tales from the *Arabian Nights* and *Hatim Tai* were available in Bengali translation[9] as well as in Marathi[10] and possibly in many other Indian languages in the mid-nineteenth century. The popularity of these books persisted for a long time. As a schoolboy in the 1880s Premchand read *Tilism-i-Hoshruba* and *Hatim Tai* with as much relish as the 'Urdu translation of the scores of volumes of Reynolds' *Mysteries of the Court of London* and the puranas of which the Naval Kishore Press of Lucknow had published Urdu translations.'[11] The sensational and spectacular elements in these stories, particularly the description of royal splendour and the extravagance of sensual delights, could not but have left their mark on subsequent narrative traditions, making it difficult today to sift the historical from the legendary in fiction.

Sanskrit literary tradition had prepared the ground for such a narrative amalgam. The word *itihasa* may have been borrowed from it to stand for history, but presentation of facts was never accorded a high premium in itihasa compositions. In Sanskrit poetics, itihasa was considered a genre of composition, like *kavya* (poetry) or *natya* (drama). Irawati Karve has observed that although in English commentary both *Ramayana* and *Mahabharata* are referred to as epics, a clear distinction was made in Sanskrit between these works, the first being kavya and the second itihasa.[12] The lexical meaning of the Sanskrit word *'itihasa'* (meaning 'it was thus') emphasizes the factual, but the literary genre bearing this name straddled chronicle as well as fiction. It may be noted here that even the English word 'history' had at one time a much more flexible connotation than it does today. Among several meanings listed in the *OED*, one is: 'a relation of incidents (in early use either true or imaginary, later only of those professedly true), a narrative, tale, story.' And as late as the mid-eighteenth century in England we find novels that are entitled histories—thus, *The History of Tom Jones, A Foundling* (1749).

In Bengali the meaning of the word itihasa had not acquired its present sense of history even as late as the nineteenth century. In the first decade of the nineteenth century William Carey published in Bengali a collection of imaginary tales and called it *Itihasa-mala*, sub-titling it 'A Garland of History'. Another such anthology published for use by the Fort William College, *Rajaboli* (1808) by Mrityunjoy Tarkalankar, made no distinction between mythological stories and verifiable events. It included episodes from the lives of Hindu, Muslim and British heroes who had fought on battlefields ranging from Kurukshetra to Plassey. These stories were, however, used as tools for language teaching rather than for teaching Indian history. But the fact remains that the popular stories of the *Arabian Nights*, when retold in Bengali, were called *Arabya Itihasa*. Similarly other stories from the Persian were published in Bengali first in 1834 as history—*Parasya Itihasa* by Nilmoni Basak—but the same book was reprinted in 1856 as fiction—*Parasya Upanyasa*.[13] This change from itihasa (history) to upanyasa (novel) may indicate a shift in the Bengali meaning of the word itihasa in the intervening period, and also an awareness of its clear distinction from narrative fiction.

This change might have come about in a number of different ways. Interest in the past is a necessary offshoot of man's active involvement in the present as something real and important (not *maya*). The new generation of English-educated men in the three metropolitan centres in India began to stake their claims in the present and this was a necessary preparation towards an interest in and study of the past. Writing of history books, specially for educational purposes—an activity not known in the earlier centuries—had begun in Bengali and in the list compiled by Bijitkumar Dutta (included in his *Bangla Sahitye Aitihasik Upanyas)* of the history textbooks written in Bengali during the first half of the nineteenth century, about one half were histories of Bengal and the other half included India and the world. The post Macaulay generation, which was exposed to English writing, had learnt to distinguish between fiction and chronicle. Bhudeb Mukhopadhyay for example was one writer who produced historical novels as well as textbooks of history, but he was fully aware of the difference between the two. Perhaps his volume called *Aitihasik Upanyasa* (1857) which contained two long stories was the first attempt in Bengali to create a new genre of fiction which used history as background, and very soon this was to be the most popular form of fiction in many parts of India.

The influence of Sir Walter Scott is often cited as important in shaping the historical novel in India, specially the works of Bankim and Romeshchandra Dutt. On a closer examination Scott's influence turns out to be much less than is rumoured, but perhaps Scott acted as a catalyst for the feeling of nationalism that was germinating in the country. The question of influence at that period is a rather complicated one. Bankim was translated into Hindi under the supervision of Bharatendu Harishchandra and towards the end of the century Kishorilal Goswami wrote a novel professedly 'under the influence of Bengali'. Although he did not mention the title of his model, it is clear that Goswami's *Sukha Sharbari* (1894) was inspired, at least partially, by Bankim's *Kapalkundala* (1866). Similarly Bankim was translated into Marathi and a writer like Harinarayan Apte might have been influenced by his work. When Apte influenced writers of the neighbouring language Kannada at the beginning of the twentieth century, it is possible that Bankim's influence filtered indirectly through him to Galagnath and Venkatachar. Venkatachar also translated some of Bankim's novels directly into Kannada. While writers of Bengali, Hindi, Marathi and Kannada might all have been influenced by Scott directly from English, indirect influence through other Indian languages cannot be ruled out. Moreover outside influences were constantly being modified by indi-

genous oral or written narrative traditions, and history often
had a regional bias, incorporating Tipu Sultan in Kannada,
Shivaji in Marathi, Pratapaditya or Sirajuddoulah in Bengali—
although Maratha and Rajput history transcended regional
boundaries to gain an all-India popularity. Local records of
history like the *bakhar* in Maharashtra and the *buranji* in Assam
proved rich quarry for fictional material. Rajanikant Bardoloi,
for example, wrote his best work *Monomati* (Assamese; 1900)
unfolding a love story against the background of the declining
years of Ahom rule, the Burmese invasion of Assam and the
consequent insecurity and unrest. In Oriya Fakirmohan
Senapati brought to life the horrors of Maratha raids in Orissa in
the eighteenth century in his novel *Lachhma* (1914).

But in many languages the most popular themes centred
around Shivaji and the Rajput kings who successfully resisted
Mughal power. The publication of English books on Indian
history made it possible for writers in any language to use
historical material from other parts of India and create a new
myth of valour. The three books most important in this
connection are Tod's *Annals and Antiquities of Rajasthan*
(2 vols, 1829), Grant Duff's *The History of the Marathas*
(3 vols, 1826), and Caunter's *Romance of History : India*
(2 vols). Tod not only wrote history, he collected fables, folk
tales and legends of Rajasthan, and even songs and stories like
Prithviraj Raso provided him with material. Many Bengali
novelists including Bankim received their inspiration from Tod
and some of the historical inaccuracies in his *Rajsingha* can be
traced back to Tod. Romeshchandra Dutt, an ICS officer whose
best historical novel is based on Maratha history, acknowledges
his debt to Grant Duff thus:

I remember the solitary evening when I encamped in the midst of the
rice-fields of Dakhin Shahbazpur, a sea washed village in the mouth of
the Ganges, when I read Grant Duff's inspiring work on the History of
Marattas and spent my nights in dream over the story of Shivaji.[14]

Shivaji provided a necessary symbol of Hindu unity, a
reaffirmation of Indian valour; and hence not only Marathi
writers but novelists from other parts of India (which was then
a defeated nation) harked back to this period of recent history in

quest of self-esteem. Rajput history was another such period of glory, and Tagore underlines the importance of this area of Indian history in an essay:

We had to learn with dates, the detailed account of how India has always been defeated and humiliated by other races from the time of Alexander to the time of Clive. In this desert of shame the only oases were the stories of Rajput valour. Everyone knows how in those days Bengali novel and drama eagerly milked every drop of inspiration that Tod's *Annals of Rajesthan* could provide. This reveals the extent to which we were starved of something that would sustain our self esteem.[15]

There is of course a special Bengali dimension to this lamentation. Unlike Rajputana or Maharashtra, the political history of Bengal after the eleventh century was of unbroken Muslim rule until it was displaced in the latter half of the eighteenth century by British control. The Hindu Bengali memory could not hark back, as the Rajput or Maratha memory could, to historical events or personages it could proudly remember. The Bengali romance with history contains therefore an element which did not enter historical fiction in other Indian languages.

With or without the support of history, the need of the Indian novelist of that period to go back to the past had one general dimension that has been candidly stated by Naro Sadashiv Risbud, author of the very popular Marathi novel *Manjughosha* (1867). He observed with regret that there was nothing in the dull present experience of Indian people which could provide the raw material of novels; the daily life of the ordinary nineteenth-century Hindu was, according to him, far too colourless to hold the attention of readers. Hence the need to choose a theme so remote that one could get away from reality.[16]

The popularity of the historical and pseudo-historical novels of the late nineteenth (and in some cases continuing well into the twentieth) century can thus be seen from three or four different perspectives. First, contact with European literature suddenly opened out for the educated Indian what seemed to him a whole new world of imagination, humanism and triumph of self over hierarchical society. That these countries too had their internal oppressions based on class and gender was not always apparent to the Indian reader. In response to this seemingly expansive world life for the middle-class Indian in the nineteenth century seemed limited, hedged in by social restrictions and politically servile. Therefore the Indian creative writer often turned to a more expansive past where human beings seemed to be of a larger stature, where valour and heroism counted, and where glory and splendour seemed infinite. Second, the so-called historical novel could be fitted more easily into the traditional concept of storytelling than realistic fiction of the western variety. *Daastan, kissa, tilism* were popular forms of narrative and the cycles of legends borrowed and adapted from Persian were available in most Indian languages. These stories usually dealt with adventure, chivalry, magic and love, and were dominated by heroes of invincible courage. Such tales allowed freedom to the imagination—in the fabled description of riches, passion and regal splendour. Moreover, adaptation of episodes from the epic and the puranas was an accepted literary practice in the modern Indian languages, and these too often emphasized heroism and valour. It is important to remember that even among Indians of the nineteenth century who knew English, realist novelists like George Eliot and Elizabeth Gaskell were never as popular as a historical novelist like Scott or romancers like Wilkie Collins, Bulwer Lytton and others who dished out romance and fantasy in different guises in multi-volume tomes like *The Mysteries of the Court of London*. Third, any past, historical or otherwise, was better than the miserable present, and the wonder-evoking though unreal happenings of a bygone era were an anodyne to the miseries of present existence. The commercial film in India today manages to perform the same function by portraying a present that is unreal but satisfying. Finally, the framework of history afforded the novelist a way to glorify the past, and the past, however nebulous, meant the pre-British past: any tale of past bravery or heroism vindicated present servitude. This was the safest form a newly awakened nationalism could take.

While almost all the pseudo-historical novels of the period will illustrate the first three points, the fourth point will be further explored in this chapter by analysing a representative and influential novel of the period, Bankim's *Anandamath* (Bengali; 1882), which was a part of the author's concerted attempt to remind Bengalis that their past was not one continuous dark night.

II

While Rabindranath in the next generation was fascinated by the history of three different groups of people—the Marathas, the Rajputs and the Sikhs, each valiant and heroic in a different way—Bankim after drawing upon Rajput history in *Rajsingha* looked wistfully backwards into the past of Bengal to find inspiring events on which to base his novels. But the recorded history of Bengal of past centuries was one of defeat, surrender and exploitation. Bankim wanted to create a new myth for Bengal in order to shake people out of their somnolence, and he found in the historical record of the Sanyasi Rebellion a convenient point where elements of bravery, religion and nationalism would be made to converge. Thus *Anandamath* was a deliberate attempt to mythicize history.

First serialized in the Bengali monthly *Bangadarshan* during 1881–2, *Anandamath* was published in book form in 1882 and gained unprecedented popularity. Five editions appeared before the century was over and by the early years of the next century it was translated into the major languages of India, in some languages more than once.[17] There were two translations in Marathi, three in Telugu, and of the three available translations in English one was done jointly by Sri Aurobindo and Barindrakumar Ghosh.[18] Some translations have appeared as late as the sixties of this century, testifying to the continuing popularity of the novel at a certain level.

Beginning with *Durgeshnandini* (1865), Bankim had written fourteen novels, of which at least seven had historical backgrounds of one kind or another. Three of these—*Chandrashekhar* (1875), *Anandamath* (1882) and *Debi Chaudhurani* (1883)—were set in the late eighteenth century during the time of anarchy in Bengal when the Muslim nawab was virtually out of power and British rule had not yet been regularized.

Anandamath is by no means Bankim's best novel, nor is its historical framework authentic, yet the novel is significant for many extra-literary reasons, specially for the tremendous impact it had on subsequent nationalist movements in Bengal and other parts of India. 'The secret societies modelled themselves closely upon the society of the "children" (*santan*) of *Anandamath*. "Bande Mataram", the battle cry of the "santan", became the war cry not only of the revolutionary societies but the whole of nationalist Bengal,' wrote Lord Ronaldshay in 1925.[19] A Tamil critic writing recently about the appeal of Bankim recognized in him 'a national purpose' which transcended regional loyalty. According to him Bankim was 'attempting in the wake of the new awakening of the country a kind of idealistic romanticized regeneration of the Hindu ethos'.[20]

It is doubtful if Bankim's vision of the regenerated Hindu ethos included the whole of India. He was more immediately concerned with the condition of the servile Bengali people. He felt that Bengalis must become aware of their past greatness so that they could rise again. In a *Bangadarshan* article of 1882-3 entitled 'Bangalar Itihas Sambandhe Kayekti Katha' (Some Thoughts on the History of Bengal) he wrote:

The nation which has memories of former greatness can preserve that greatness, or if lost, try to restore it. Blenheim and Waterloo are the results of the memories of Crecy and Agincourt. Italy has been able to rise again after her fall. Today the Bengalis want to be great. Alas! where is their historical memory?

We need a history of Bengal. Without this Bengalis can never rise to their full stature.... Bengalis who are convinced that their ancestors were always feeble and without substance, that their predecessors never achieved anything glorious—such Bengalis cannot aspire to any other than an unsubstantial, inglorious condition, nor would they strive for anything different....

And where would we find substance? Has a proper history of Bengal been written yet? Englishmen have written numerous histories of Bengal. Stuart Saheb's book is so large and heavy that it would kill a strong young man if thrown at him. Marshman, Lethbridge and others have made a lot of money by writing slighter versions of Bengal history.

But do these books tell us anything historical about Bengal? In our considered opinion, not one of these English books contains the true history of Bengal.[21]

Through his novels as well as his other prose Bankim sought to remind the Bengali people that they had a worthwhile past. *Anandamath* was very much a part of this lifelong endeavour. But even though the national identity and past glory of India as a whole were not Bankim's direct concern, *Anandamath* eventually came to exercise a pan-Indian appeal at a certain level. This happened because this novel fused among the Hindus a revived religious fervour with a new-found patriotic zeal. Here, perhaps for the first time in Indian history, the concept of the mother goddess with its connotation of shakti became linked with the idea of the nation as a political unit. The symbolic power of this fusion proved so far-reaching that the consequences are visible even today.

The action of the novel is placed in the 1770s, a decade which saw in Bengal a dreadful famine as well as the so-called Sanyasi Rebellion. Although the famine and the insurrection are historically recorded events, Bankim did not emphasize the historicity of the novel in the first edition. In the second edition (1884) he added, in response to popular demand, an intro-duction where he quoted from Gleig's *Memories of the Life of Warren Hastings* and W.W. Hunter's *Annals of Rural Bengal* to show what had actually happened in 1773, but he also disclaimed that he had wanted to write a historical novel.[22] As a historical novel *Anandamath* is patently anachronistic because the 'motherland' nationalism depicted in it was nowhere present in 1773 even as an abstract idea, and the Sanyasi insurrection did not have any ideological overtone. Bankim himself admitted in one of his essays in *Bangadarshan* that the concept of nationhood was the direct result of English education.[23] The novel thus reflects the concerns of Bankim's own time, a period marked by the rise of nationalism. This newly awakened consciousness made itself felt in various public events of the time, including the agitation against the Ilbert Bill in 1883 (a year after this novel was published) and the beginning of the Indian National Congress three years later. It has been pointed out by a historian that the theme of *Anandamath* has striking similarities with the actual career of a militant nationalist Wamdeo Balwant Phadke (1845–83) who wore the robes of a sanyasi and attempted to raise an army for 'destroying the English' and who was finally arrested and convicted by the British in 1879.[24] The immediate cause which drove Phadke to take the vow of 'destroying the British' was the terrible famine of 1876–7, which ravaged western India and caused unbelievable suffering to the people.[25] The fact that the background of Bankim's *Anandamath* is the devastating famine of 1769–70, which was said to have wiped out a third of Bengal's population, and that in this novel he deals with the nationalistic endeavours of a group of sanyasis reminiscent of Phadke's group, is not a mere coincidence. Bankim was a keen reader of history and contemporary affairs and it is likely that Phadke, who was very much in the news in 1879, gave him the germ of his theme which he later incorporated conveniently into an episode from Bengal history. That he never mentioned the Phadke episode and did refer to the Sanyasi insurrection in the preface to the second edition is also explicable in view of the fact that he was a government servant: it would not have been politic on his part to glorify a recent act of sedition.

In the seventies of the eighteenth century all of Bengal had not come under British rule. The British merely collected the revenue while the welfare of the people rested in the hands of the weak and dissipated nawab. To quote from the opening chapter of *Anandamath*—

[If] Cowardly Mirjafar—the heinous traitor—was unable to protect himself, how would he protect the lives and property of the people of Bengal? Mirjafar drugged himself and dozed. The British extorted the revenue and wrote despatches. The Bengalis merely wept and resigned themselves to their ruin.[26]

The novel is set against this nightmarish misrule and a countryside stalked by famine. Mahendra Singha, a prosperous householder, is driven out of his village by the prospect of starvation. While travelling through the devastated land he is separated from his wife and child, and encounters a band of rebel sanyasis who call themselves 'santans'—the children of the Mother who organize themselves against the oppressors, hoping to bring back the pristine glory of the motherland. By gradually sapping the strength of the enemy in guerilla warfare they finally rout the combined forces of the Muslim and the

British in an open confrontation. Since this victory could not be historically authenticated, in the last chapter of the novel Bankim introduced an ascetic with prophetic vision who dissuaded the rebels from going any further. It was necessary, he advised them, to submit to British rule until the true religion of the Hindus could be purified from its present degenerate and corrupt state by new empirical and scientific knowledge. Intertwined with this central account are several human stories highlighting the tension between the santan's vow of renunciation and the normal human instincts.

The closure of *Anandamath* draws our attention to the author's attempt to negotiate between the evidential and ideological levels. The importance of the book, lies not in the authentic portrayal of a historical period but in its impact on the people who read it. It consolidated the inchoate ideals and aspirations of a people who needed a new myth. And the ending of the novel interests the reader today as an example of the characteristic ambivalence of Bankim's as well as most educated Indians' attitude towards British rule, an attitude which wavered between the dream of an independent India and admiration for the British. Bankim acknowledges the fairness and the justice of the British in the very novel in which he inspires the Bengali people in sonorous rhetoric to stand up against alien domination. Not alone in internalising this duality, Bankim nevertheless provides a convenient site for investigating this syndrome common among the colonial elite. When the novel emerged as a genre in India, British rule was firmly entrenched in Bengal. The mutiny of 1857, which had such strong reverberations elsewhere in northern and central India, did not cause much stir in Bengal, where there was generally a feeling of gratitude among the educated class towards the British for bringing about law and order in a trouble-torn land. The following passage from *Anandamath* describing conditions in a previous century indirectly illustrates this feeling:

North Bengal was no longer under Muslim control. The Muslim rulers did not want to recognize this fact. They deluded themselves by saying that a few bandits were causing sporadic trouble. But by God's grace Warren Hastings became the Governor-General at Calcutta. He was not the kind of person who would delude himself. Had he been such a person where would the British empire be today?[27]

Bankim served the British administration as a Deputy Magistrate continuously for a period of more than thirty-two years without a single promotion throughout his career—except for the few months when he was appointed Assistant Secretary to the Bengal Government in September 1881. This post was a temporary one, but when he was reverted from it there was a general protest in newspapers which included the British-owned *The Statesman*. Whether there was any connection between this event and the fact that *Anandamath* was being serialized in *Bangadarshan* during this period cannot be conclusively proved.[28] But it is not difficult to guess that the rousing nationalism of the novel could not have pleased the rulers. The press had been effectively gagged by Lord Lytton and there is evidence that Bankim published the first edition 'after a good deal of deliberation'.[29] He is said to have expressed a desire to write a book on the Rani of Jhansi, but refrained from doing so because he had already offended British officialdom by writing *Anandamath*.[30]

Recent research on Bankim in Bengali has revealed that some of the passages in *Anandamath* which are adulatory of the British were later additions, possibly the result of diplomatic afterthought. By comparing the five editions of *Anandamath* which appeared during Bankim's lifetime with the original serialized version in *Bangadarshan* it can be shown that Bankim was continuously attempting to save the situation by adding or subtracting words and sentences to make the book less offensive to the British.[31] For example he frequently softened the adjectives applied to the British as enemy, or substituted the word British (*Ingrej* in Bengali) by *yavana*. In the novel both the Muslims and the British are enemies and the words *yavana* or *bidharmi* (i.e. those belonging to a different religion) can refer to either. The country has to be saved from both of them, but the author sometimes exploits the ambiguity of the word and remembers to comment on the relative superiority of the British as a race. Considering Bankim's official position it is not easy to decide today how much of his admiration of the British came out of conviction and how much out of expediency.

Bankim has been taken to task by modern critics for his Hindu chauvinism and his consequent prejudice against Muslims.[32] Bankim himself was not unaware of this possible allegation by posterity because he added a rather gratuitous and defensive postscript to his novel *Rajsingha*, stating that in this book he had by no means tried to indicate the superiority of Hindus over Muslims. He attempted to give an academic colouring to his denuciation of Aurangzeb by comparing him to Philip II of Spain. Possessing great qualities in themselves, both Philip II and Aurangzeb, said Bankim, were destroyed by certain flaws inherent in their characters and were humiliated by enemies less well-equipped than themselves—by the British and the Dutch in one case and the Rajputs and the Marathas in the other. The Maratha warrior Shivaji, according to him, was comparable to Queen Elizabeth of Britain, and Raj Singha to William the Dutch.[33]

This tactful and astute statement is another example of the ambivalence mentioned earlier. Glorifying Shivaji, a potentially seditions act, is neutralized by equating him with an earlier queen of the ruling race. And while one can by no means ignore or whitewash the fact that in *Anandamath* Bankim's concept of India was a purely Hindu nation where patriotic ardour could be subsumed into a devotional zeal and national regeneration could be identified with revitalization of religion,[34] elsewhere his chauvinism seems tempered by an understanding of the economic reality of the country. In the long essay called 'Bangadesher Krishak' (The Peasants of Bengal) he lashes out against the exploitation that both Hindu and Muslim peasants have to suffer. The author who in *Anandamath* seemed grateful to the British for bringing law and order to the land, here castigated the Permanent Settlement Act of Lord Cornwallis thus: 'The blame for this state of the peasants in Bengal falls squarely on the British. Their Permanent Settlement is a permanent arrangement for ruination and they will never live down this permanent shame.'[35] One explanation of this inconsistency in Bankim's attitude towards Muslims could be that he regarded the Muslim peasant as Bengalis, whereas the Muslim rulers of Bengal were to him an alien people.

His undisguised hostility towards the yavana or the bidharmi rulers is not so much a stricture against their religion as against their status as foreigners. When Bankim talked of the present he could see the Hindu and the Muslim both as part of the exploited people. But in the context of the past his image of the Muslim changed into the alien oppressor and his position automatically shifted.[36]

A series of antinomies runs through the entire career of Bankim. This so-called feudal and communal bourgeois writer understood the plight of the exploited peasant and pleaded for the greater good of the common man.[37] The novelist who sympathized with the anguish and deprivation of the child-widow (Rohini in *Krishnakanter Will*) also felt the need to kill her as a punishment for threatening the stability of the social order. One who was a faithful servant of the British also made fun of the English officers in *Anandamath* by depicting them as comic and venal figures, and enabled the Hindus to vanquish them. The supposedly Hindu revivalist writer propagated in *Anandamath* a religion that was very different from any recognizable form of Hinduism. It was a unique synthesis of the Vaishnava cult with the Shakti cult, something that despite the popularity of the novel was never accepted by Hinduism. Thus Bankim's religious thought here has a certain non-traditional quality and may even be a fictional strategy.

Anandamath is undoubtedly flawed as a novel. It is even said to have marked the beginning of 'the decline of Bankim's power as a novelist.'[38] The influence of the novel is usually strongest on the adolescent mind, and later with greater ideological and aesthetic sophistication one tends to grow slightly embarrassed at the hold it once exerted. Yet reading the novel a hundred years after its publication one finds that it is more than a simplistic work of Hindu propaganda. There is a complexity and an ambiguity contained in its historical and political substrata that tell us more about the contradictions involved in the situation of the nineteenth-century educated Indian—and his attitudes towards the past and the present—than any factual social report can.

III

Bankim wrote in the Introduction to his next novel, *Debi Chaudhurani* (1883), 'I shall be much obliged if my reader does not consider *Anandamath* or *Debi Chaudhurani* as historical novels'—even though the adventures of the woman bandit described in the novel did have an analogue with something similar contained in a report on the Rangpur district of Bengal and included in Hunter's *Statistical Account.*[39] Bankim had said earlier, in his Introduction to *Anandamath*: 'The Sanyasi Rebellion is a historical fact but I do not see any need to tell the reader about this.' On the title page of his first novel, *Durgeshnandini* (1866), he says that it was an '*itibritta-mulak upanyasa*'—i.e. a 'novel based on chronicle'; his third novel, *Mrinalini* (1869), he introduced as a historical novel. But in the introduction to *Rajsingha* (1882) he stated, 'I have never written a historical novel before this. *Durgeshnandini, Chandrasekhar* or *Sitaram* cannot be called historical novels',[40] and from the third edition *Mrinalini* ceases to be called a historical novel. Thus Bankim's own idea of a historical novel was constantly undergoing a change. His basic objective, however, did not change much. Not the accurate recreation of a historical period but awakening the national pride of his readers was his primary intention. One of his reasons for starting the monthly Bengali journal *Bangadarshan* in 1872 was to create an atmosphere of self-respect among educated Bengalis through the study and discussion of their past. *Bangadarshan* must have had an adequate impact on its readers because in the next generation it became possible to start a journal devoted only to history. Named *Aitahasik Chitra*, it began in 1898 and declared its purpose thus:

Our intention is to publish in translation from other languages, accounts of foreign travellers to India, standard books of history, results of historical research, reviews of modern history books, and the family histories of landowning families and their old chronicles.[41]

Though not with the same intention, some scholars in Maharashtra had moved in a similar direction. Attempts made by Nilkantha Rao Kirtane, Govind Vishnu Ranade, Chiplunkar and others to refute Grant Duff's view of Shivaji culminated in the founding in 1878 of the journal *Kavyetihas Samgraha*, which, in the opinion of the historian D.V. Potdar, marks the starting point of real historical research in Maharashtra.[42]

Though known hitherto mainly as a poet, Rabindranath wrote for the first issue of *Aitahasik Chitra* and contributed a few reviews subsequently. He was very enthusiastic about the enterprise and speculated on the ways and means by which this newly awakened interest in history could be spread beyond the educated elite to the masses. Since the two most popular forms of entertainment in rural Bengal were the *jatra* (folk theatre) and *kathakata* (oral recital of the purana stories), Rabindranath suggested that historical themes could be introduced there along with the standard purana tales: 'People could be easily entertained by the stories of Prithviraj, Guru Govinda, Shivaji and Akbar. And moreover related in the voice of a good kathak there is no reason why the stories of *Rajsingha* or *Anandamath* should also not be enjoyable to the people'.[43]

Rabindranath's choice of the example of historic figures in the above passage indicates his catholicity of vision which takes in different regions and religions. He was indeed the first visionary to create a national consciousness of India in literature. If Romesh Chandra Dutt and Bankim used Rajput or Maratha stories, they did so because not much historical material was available in Bengal. Bankim's real concern was with the identity of the Bengali people, and in order to arouse them he did not hesitate to resurrect rumours of a colonizing past. Citing with approval Sir George Campbell's remark that Bengalis were the Athenians of Asia, Bankim wrote in the *Bangadarshan* of 1873–4:

Truly, if not in any other respect, Bengalis were like Athenians in their colonial enterprise. Simhala was conquered by Bengalis and occupied by them for several generations. It has also been conjectured that the islands of Java and Bali were Bengal's colonies at one time. Tamralipte was once a port from which Indians set out on voyages. No other race in India has shown such colonizing prowess.[44]

This may be contrasted with the seemingly wider perspective in which Bharatendu Harishchandra (1850–85) spoke of his country at a lecture he gave at Balia in 1877: 'He who inhabits Hindustan, whatever be his colour or caste, is a Hindu. Help the Hindu. Bengali, Maratha, Panjabi, Madrasi, Vedic, Jain, Brahmo, Musalman, all should hold one another's hands.'[45] It should be noted that while the lessons from history learnt by Bharatendu make him posit a greater national identity of the Indian people as a whole, there is an ambivalence in his wanting to merge this with a 'Hindu' identity—more so when we recall his clarion call of 'Hindi–Hindu–Hindustan.' When we turn to Rabindranath we realize that it was he who first saw in the newly awakened sense of history an expansion not only in Indian time-consciousness but also in space-consciousness:

Just as we want Bombay, Madras and Punjab to be near us, similarly we want a direct knowledge of the past of India. Fully aware of ourselves we want to understand our identity both in space and time, as a unified and great nation.[46]

After the false start he made with his first novel *Bau-thakuranir Haat* (1883), Rabindranath never again attempted a historical novel himself. But in 1898 he wrote a perceptive essay on the historical novel as a genre. In this essay one sees the culmination of the sense of history, the seeds of which Bankim had so assiduously planted a generation ago. Rabindranath challenges the opinions of Freeman and Palgrave—who had stated, 'Those who want to know about the age of the Crusades should on no account read *Ivanhoe*'[47]—that the historical novel is the enemy of history as well as of fiction. Rabindranath emphasizes that the primary purpose of historical fiction is not the accurate recording of facts but the evocation of an atmosphere larger than that of our ordinary daily life, reminding us of the magnificent rhythm of the chariot of time:

The joys and sorrows of ordinary men are enough for him—enough to cast in shadow the larger events of the world. .... But once in a while a few individuals appear whose joys and sorrows are linked with the greater rhythm of world events and the rise and fall of kingdoms. Their private loves and feuds merge with the great orchestra that like the roaring of the oceans swells and subsides according to the large and remote workings of Time.

...The link between such men and the movement of history is not perceptible to our limited quotidian view. If, per chance, a great man, a maker of history, is living in our midst, we cannot perceive him and the larger span of history simultaneously through our fragmented vision. . . . In order to see such men not as individuals but as a part of history we have to stand back, place them in time so that we see them along with the enormous stage where they were the protagonists.

This distancing from our ordinary lives is important. While we are spending our days in routine jobs, in laughter and tears, eating and sleeping, in the broad thoroughfares of the world the chariot of time is being driven by men larger than us. This realization gives us a momentary release from our circumscribed existence. This is the true aesthetic experience of history.[48]

He goes on to suggest that the aesthetic enjoyment generated by the historical novel is so different in kind from literature of other varieties that one could almost add a tenth *rasa* to the nine formulated in Sanskrit poetics and call it the *aitihasik rasa*. The question of 'historical truth in the artistic reflection of reality,' which is Lukacs' concern in his *The Historical Novel*,[49] is perhaps another version of this search for a new rasa. According to Lukacs the eighteenth century realists (Lesage, Defoe, Fielding) who accomplished a major 'breakthrough to reality' in fiction portrayed their contemporary world with unusual plasticity, but they did not see the specific qualities of their own age historically. Even in Europe, for the realization of historical truth in fiction one had to wait till the beginning of the nineteenth century.

IV

Novels that drew their material from history were slow to emerge in Hindi. When at the turn of the century Kishorilal Goswami and Gangaprasad Gupta began using the past in their fiction, their intentions were clearly different from those of Bankim and Romesh Chandra Dutt. In the seventies and early eighties Bharatendu had done a great deal to create the right atmosphere for the writing of historical fiction in Hindi, but in his lifetime (1850–85) no original novel of this kind appeared. What we have instead are translations, mainly of Bengali novels. Gadadhar Singh, who had translated Romesh Chandra Dutt's *Bangabijeta* at Bharatendu's request, began translating *Durgeshnandini* in the late seventies. This, after being serialized in *Kavivachansudha,* was published in two volumes in 1882 and 1884. Bharatendu had also arranged to get *Rajsingha* translated into Hindi; he wrote the entire first chapter, besides revising some other sections. This was published in 1894, long after his death, by Khadagvilas Press of Patna. Radhacharan Goswami translated another historical novel from Bengali— Swarnakumari Debi's *Deep Nirban*— in 1891, and Romesh Chandra Dutt's *Maharashtra Jiban Prabhat* appeared in Hindi in 1881.

These translations do not so much cater to popular taste as reflect the enthusiasm of a few dedicated and idealistic writers, generally known as the Bharatendu group, who wanted to serve the cause of Hindi by enriching it from various sources. When Bankim was told that *Durgeshnandini* was being serialized in Hindi he is said to have asked for his share of the profits if it appeared as a book. He had no idea that a writer of serious fiction in Hindi could not expect any profit in those days. Hindi literature was merely a cause to uphold. Gopal Rai, in his extended work on fiction and the Hindi reading public, has shown by quoting publication figures that none of these books ever had a second edition: 'The readership that did exist tended to read *kissas* and *kahanis.* Urdu dominated the scene and Hindi continued to be neglected as a literary language'.[50]

The efforts of the Bharatendu group to evolve a literary ethos in Hindi began to bear fruit towards the end of the century. Other public events like the establishment of the Indian National Congress in 1885, the founding of the Arya Samaj in 1875 and the gradual spread of English education in North India, all had an indirect impact on Hindi-reading people, arousing in them an awareness of nationalism and consequent interest in the past. The fact that novels based on the historical past came to be written in Hindi at the end of the nineteenth century and became popular only in the early twentieth century—whereas historical novels had taken roots in Marathi and Bengali more than a generation ago—has to be seen in the context of the regional variations in social conditions and educational and economic opportunities in nineteenth-century India. Not the least important among these complex variables was the fact that the three first western-model universities were set up in India in 1858 in Calcutta, Bombay and Madras, three cities where British influence had already been concentrated. Dissemination of western thought and absorption of English influence which led to a historical and political consciousness had thus happened earlier around these coastal cities than in inland regions.

The first generation of popular novelists in Hindi—Devakinandan Khatri, Kishorilal Goswami and Gangaprasad Gupta—who started writing in the eighteen-nineties generally placed their novels in the past, but this was the unspecified hazy past of romance. Khatri, particularly, evolved a totally new kind of past-based fiction that had no relationship with history. The other two used historical settings but did not confine themselves to historical events or credible human achievements. Kishorilal Goswami's *Hriday Harini* (1890), *Lavangalata* (1890), *Tara* (1902) and *Razia Begum* (Part I, 1904 and Part II, 1905) are set mostly in the courts of Muslim or Rajput kings or of the Peshwas, but it is not easy to disentangle the strands of history from the skeins of fantasy because magic and the supernatural play an important part in these narratives. Gangaprasad Gupta, whose historical novels include *Noorjehan* (1902), *Veer Patni* or *Rani Sanjogita* (1903), *Kunwar Singh Senapati* (1903), *Puna Mein Halchal* (1903) was, like his contemporary Kishorilal Goswami, a staunch Hindu. In any delineation of Rajput–Muslim or Maratha–Mughal conflict their primary concern was glorification of the superior moral values of the Hindus. Muslim characters were shown to be lecherous and cruel, indulging in a life of luxury, while the Hindu characters generally appeared upright and honest. The charge of Hindu chauvinism which has been brought against Bankim can also be brought against most Hindi writers of this period. These novelists, however, were not unique in their attitudes. Slogans like 'Hindi–Hindu–Hindustan' remind us that in the formative phase of Indian nationalism concern for Hindi as a language different from Urdu, concern for the identity of the Hindus as a religious group, and the idea of India as a nation tended to become synonymous in parts of North India. Hindu-

Muslim unity might have been the formulated policy of the Indian National Congress, but on an emotional level the appeal of religion was greater than that of secular nationalism. This may not have been equally true in all the languages, because we are told: '... the earliest Oriya historical novels are remarkably secular. They depict the suffering of the Oriya people caused both by the Muslim and Hindu invasions and seek to project the periods of crisis in the history of Orissa.'[51]

The tenuous nature of Kishorilal Goswami's interest in history can be seen if his novel *Tara*, subtitled *Kshatra Kula Kamalini*, is compared to Bankim's *Rajsingha*, both of which have the same Rajput king as the central figure. The germ of Bankim's novel came from a paragraph in Tod's *Annals* which refers to the princess of Rupnagar, a small state seeking the protection of Raj Singh, the Rana of Udaipur, against the assault of Aurangzeb.[52] But in the preface to the fourth edition of *Rajsingha* Bankim scrupulously acknowledged his other sources also and explained which facts were taken from Tod and which from Orme. He insisted that there was historical evidence to prove that life in the court of Aurangzeb was indeed licentious and corrupt, even if the king was not so. He even offered to prove it with evidence if anyone challenged him.[53] And yet the intensity with which he presents Aurangzeb's daughter Zebunnisa as a wanton woman must surely have been his own. When Mubarak, a commoner who loved her, wanted to marry her, Zebunisa scoffed at him:

Am I the daughter of a Hindu brahman or of a Rajput that I should be married to one man and remain his slave all my life, finally to die on his funeral pyre? If Allah wanted this fate for me he would never have created me as a princess.[54]

This pride in being a princess was vanquished at the end of the novel. When a transformed Zebunissa begged to be Mubarak's wife she realized that her emotional needs were the same as those of any common woman. Although highly sensational, verging on the prurient, this strand is not entirely extraneous to the structure because it offers a counterpoint to the main plot. Just as Zebunissa, intoxicated by wealth and power, spurned the proffered love of a sincere man only to repent later,

Aurangzeb himself, drunk with power, became oblivious of the people and arrived at a crisis which shook him out of his complacence. Rajsingha stands out in contrast as a steadfast, loyal and noble character.

Except for the strongly negative attitude towards the Muslims there is very little in common between *Tara* and *Rajsingha*. If Bankim felt the need to cite historical sources in order to silence the suspicious reader, Goswami disarmed his reader by disclaiming all historicity. In the Introduction to *Tara* he said he had no faith in histories written by Muslim historians: 'That is why in my novels I have focussed primarily on my imagination and history has been made secondary. At places imagination has only saluted history from a distance'.[55] In Goswami's novel thus, Raj Singh's period is freely changed to coexist with Shahjahan's regime. Unlike the princess of Rupnagar as described by Tod and Bankim, Goswami's heroine Tara has grown up in the atmosphere of the Mughal court after her father Amar Singh brought her there as a child. She sends an appeal to Rajsingh to rescue her from the royal palace in Agra. Tara and her friend Rambha are capable of miraculous feats by which they protect themselves from the amorous courtiers and their intrigues. All the Muslim characters are seen to be venal and cruel, including Shahjahan's eldest son Dara Shikoh who has been presented by most historians as a very different kind of person altogether.

If this communal bias becomes an unstated background of the historical novel in Bengali and Hindi (and also Marathi), what did contemporary Urdu novels offer, specially those written by Muslim authors? Although at that time most educated men in North India read and wrote Urdu for practical and public purposes, the bifurcation of the two languages, Hindi and Urdu, along commual lines had begun to be evident in creative literature. Historical romance was a popular genre in Urdu fiction as well. Abdul Halim Sharar (1860–1926) was a pioneer in this field and his novels, published between 1887 and 1907, enjoyed a wide popularity. Ralph Russell describes these works thus: 'Stated in the broadest term they have a single theme—the portrayal of the glorious past of Islam and the great superiority of Islamic civilization in its heyday over that of contemporary non-Muslim (specially Christian) powers'.[56] It is interesting to note that while in *Tara* and *Rajsingha* excessive sexual appetite and religious fanaticism are seen as the dominant traits of Muslim characters, in Sharar's work, specially in *Flora Florinda* (1893), the same qualities are attributed to the Christian characters. *Flora Florinda*, one of the most popular of Sharar's books, was situated far away in time and space—in ninth century Spain. The action-packed thrilling tale of love and lust, disguise, adventure and violence set against the background of rocks, castles monasteries and dungeons, in spite of shades of the gothic novel, was more in the tradition of *daastan* than of the novel. Faiz Ahmad Faiz's analysis of this nostalgic glorification of an imagined past might with slight modification apply to both Urdu and Hindi novels of the period:

Sharar's ... is an age when the Muslims had just awoken to a consciousness of their decline. These romantic tales in the first place helped them to forget the bitterness of everyday life. Secondly, the recital of past conquest partly inspired them with self-respect and partly with emotional solace, with the thought that even if *they* were not heroes, at least their forefathers had been. And thirdly, the description of the vices of other peoples provided them with a way of taking mental revenge for their present subjection ... This is why Sharar's novels were so popular.[57]

A special point to be noted here is that while Hindu memories of past glory seldom went beyond the boundaries of Hindustan, the Muslim evocation of a glorious past could hark back to the days of Moorish domination of Spain and other Mediterranean lands. Both these voyages of nostalgia – one only in time and the other combining time and space – served the novelists well and helped to enlarge the audience of the novel in India in the early years of its emergence.

V

In the last decade of the nineteenth century the readership of Hindi and Urdu fiction was not sharply divided. The action, suspense and sensation that made Sharar popular were also the elements behind the phenomenal success of the famous six thousand page trilogy by Devakinandan Khatri: *Chandrakanta*

(1891), *Chandrakanta Santati* and *Bhutnath* (the third volume was completed by Devakinandan's son Durgaprasad Khatri after his father's death). Very soon *Chandrakanta* became a cult, and in ten years six editions of several thousand copies were sold out. When one takes into account the low percentage of literacy, limited reading habit and modest buying power of most Hindi readers of the period, this figure assumes impressiveness. The continuing popularity of *Chandrakanta* in the next three or four generations can be seen in the fact that between 1891, the year of its first publication, and 1961, forty-five editions had come out, making the number of printed copies 1,80,000, even by a conservative estimate. *Chandrakanta* was printed in instalments in the Lahari Press at Benares, where it is said that there was always a crowd of twenty to thirty people anxiously waiting to find out what happened next. Before the proofs could be send to Khatri, these readers would snatch a quick look to assuage their suspense.[58]

Bharatendu and his group had done a great deal for the enrichment of Hindi and the raising of the standards of literature. But before these qualitative gains could be consolidated, a quantitative expansion achieved by Devakinandan Khatri gave Hindi as a language and the novel as a genre a firmer foothold in North India. While most critics have dismissed these novels as kitsch, it is customary since the time Ramchandra Shukla acknowledged this fact to give Khatri the credit for gaining a vast readership for Hindi fiction:

In the history of Hindi literature, Babu Devakinandan Khatri will always be remembered as someone who created more readers in the Hindi language than any other writer. Countless Urdu-knowing people learnt Hindi in order to read *Chandrakanta*.[59]

Khatri had none of the idealistic intentions of the writers of the Bharatendu group; he merely wrote to entertain. His success was due partly to his simple language — a sprightly mixture of Hindi and Urdu without any literary flourish, almost the language of common speech — partly due to his ability to create suspense through complex interlocking episodes of adventure and mystery, and lastly because of that indefinable factor that makes a book a runaway bestseller—the ability to satisfy some unconscious psychological needs of the moment.

Although the central characters are royal personages, these tales of magic and deceit have very little to do with history. The phenomenal success of this trilogy started a new trend in Hindi—a category that came to be called the fiction of *tilism* (enchantment) and *aiyari*(imposture or fraud). There is no specific reference to time in any of these books but the ambience is medieval. Battles are fought as simple direct confrontations on level ground and horses remain the chief means of transport. The first chapter of the first book of *Chandrakanta* begins thus:

It is evening and there is a red glow in the sky. In a vast deserted field, at the foot of a hill two men—Virendra Singh and Tej Singh—are sitting on a rock talking to each other.

Virendra Singh is about twenty-one or twenty-two years old. He is the only son of Surendra Singh, the king of Naugarh. Tej Singh, his closest friend is the beloved son of Surendra Singh's minister Jeet Singh. Extremely clever and nimble, Tej Singh carries only a dagger at his waist and a bag hanging from his shoulder. He has a rope ladder in his hand. He is briskly looking around on all sides as he talks to the prince. In front of them stands in a state of readiness a well-equipped charger, fastened to a tree.

For several thousand pages the novel and its sequels continue in this apparently simple manner, but subtly weaving a complex fabric of enchantment in which before one knot can be unravelled several other knots begin forming. At any given moment the reader is in the midst of several mysterious tangles, each at a different stage of complication or solution. Thus the storytelling is manipulated with superb craftsmanship. Almost all the miraculous feats of the characters are given pseudo-rational explanations without once taking recourse to the supernatural. Khatri pretends to give scientific explanations for the doors that open and close on their own, the instruments that can play music unaided by human hands, the walls whose touch can make people unconscious and the statues that can laugh. His attempts were so convincing that many of his newly literate readers regarded the adventures of Virendra Singh and Tej Singh as accounts of actual events. To clarify his own position Khatri wrote with careful ambivalence:

Just as books like *Panchatantra* and *Hitopadesha* were written for the instruction of people, these are written for their entertainment ... Now remains the question of plausibility ... What was impossible a hundred years ago is now becoming possible through science. In an earlier age who could have believed in the power of electricity, trains and wireless?[60]

This veiled attempt to incorporate technology with magic is very similar to the strategies of some of the action movies of our time, both Hollywood and Bombay, where electronic and other devices are used to magnify the incredible feats of the hero or the villain, making the rational and the non-rational coexist and serve each other.

These references to applied science are the only indication in these novels that India has been exposed to the western technological world-view. There is otherwise no reference to a foreign presence. The kingdoms of Naugarh, Chunar, Vijaygarh, etc. appear to be soverign states, more self-sufficient than the Rajput states—which always had to live under the shadow of Mughal threat. The few Muslim characters that exist tend to be villainous and the general tenor, notwithstanding the display of valour and the material obsession with hidden treasures, is piously Hindu. It seems like a last ditch attempt to reassert values that seemed threatened by a stronger culture. The popularity of *Chandrakanta* may thus be attributable to the impotent nationalism of a defeated people who attempted to appropriate the new technology as part of their own ancient glory.

According to Rajendra Yadav, a Hindi novelist who has written an incisive sociological critique of this trilogy by relating it to the cultural and economic axes of India's colonial plight, *Chandrakanta* is born out of a terrible feeling of national inferiority, although it has been made palatable by several layers of wrapping. Yadav thus places Devakinandan Khatri and his magnum opus in a very central and crucial position in North Indian culture of the time:

When we look back at the point where history takes a turn we usually find an event, a movement or a person who/which embodies the pressures generated by the impact of time and the forces of society ... Sometimes, if we look carefully, we might even find a book ... At the end of the nineteenth and the beginning of the twentieth century in north India, or rather in the Hindi speaking milieu, there existed such a book—*Chandrakanta, Chandrakanta Santati* and *Bhutnath*—one book in all.

... We learn that Goethe's *The Sorrows of Young Werther* captured the imagination of German people and *Anandamath* appealed to the youth of Bengal when they first appeared, in exactly similar ways. *Devdas* is another such book. If one has to speculate on the psychology of a society on the basis of its popular books I would suggest that *Anandamath* showed the frustrated readers a way out of their anguish of inferiority through the path of action and hope. *Devdas* in a later generation came, on the other hand, as the elegy of the romantic despair of youth. It celebrated their inaction and defeatism ... I would like to suggest that *Chandrakanta* is not only the confession of defeat, but after the failure of physical power in the feudal context it is a conscious attempt to glorify the mind. In other words after the acceptance of defeat in valour it is a weak and illusory attempt to prove the superiority of the intellect and ingenuity.[61]

Yadav's forty page analysis makes it clear that not only in the case of *Chandrakanta* but also in the evaluation of the other popular novels of the nineteenth century set in the past, literary tools alone are not sufficient. Many other factors have to be considered—the co-ordinates of history, both political and economic, and of religion as well as philosophy. Reality peeps through the chinks of the marvellous, reminding us of the tensions that created these novels. These novels in turn are reminders that a great deal of truth can be revealed by means of fiction and fantasy.

IV

WOMEN IN A NEW GENRE

If placing novels in the past afforded novelists opportunities for fabulation, those who dealt with the present generally attempted to write in the realistic mode. This was not an easy task because when the new genre came into its own in India in the last quarter of the nineteenth century, urban life was undergoing changes of several kinds simultaneously. This created tensions and paradoxes unknown before. In the sphere of education alone not only had a new language become available to the Indian elite but English had also uncovered new values and opened up new modes of thought, including the ideal of individualism— an ideal that was not easily reconcilable with the hierarchical and role-oriented structure of traditional Indian society. The novelists who attempted to present this complex period in fiction were themselves products of this tension.

Since realism was the dominant mode of the British Victorian novel—the model immediately available to the Indian writer— his primary challenge was the achievement of realism while remaining faithful to the reality of a social order which generally inhibited individual choice. The problem was not to be easily solved because the evolution of the novel in England had been closely connected with the emergence of the ideal of individualism in life and the perfection of realism as a mode in fiction. These were both reflections of a basic ideological shift. Todorov is right in pointing out the historic inevitability of literary genres: 'It is not chance that the epic is possible during one era, the novel during the other (the individual hero of one being opposed to the collective hero of the former): each of these choices depends upon the ideological framework in which it operates.'[1]

As a narrative art form the novel differs from the epic on the one hand and from the romance on the other, especially in terms of characterization. The writer of romance does not attempt to create 'real people' so much as stylized figures which expand into psychological archetypes. Creating real people in a recognizable historical setting—people who are not mere archetypes or representatives of a caste or a class or a social role (priest, landlord, mother-in-law, etc.)—necessitates an acceptance of subjective individualism and a specific awareness of history. The latter had never been a component of traditional narrative in India, and the former was not easy in a tradition-bound society, even though writers themselves had begun to be restive. Changes in the writer's own value system were perceptible but these had not made any dent on the larger social structure, and to this extent the major Indian novels of the nineteenth century reflect a central dilemma of the period.

Returning to the western model once more we find that the growth of individualism in English fiction after the eighteenth century can be related, among other factors, to the new social mobility that industrialization had made possible, making man realize his unique potential outside a rigid hierarchy. Imperial expansion was another factor of mobility, as was the breakdown of hereditary professional categories. Nothing comparable had happened to nineteenth-century English-educated urban Indians who were the creators as well as the consumers of literature. Their mental horizon may have undergone a basic change, but in terms of economic potential the range had not expanded. The aspiration of most educated Indians was to find employment under the British, and thus to the hierarchy of the traditional social and family structure was now added a new colonial hierarchy.

The depiction of love and courtship was a major concern of English novelists of the eighteenth and the nineteenth centuries. Around the time the novel emerged in English, the relationship between the sexes was undergoing a change. The concept of romantic love was displacing that of courtly love and marriage was slowly becoming a matter of individual choice rather than familial obligation.[2]

In late nineteenth-century India not only were conventions of marriage restrictive, even social intercourse between the sexes was not common in the upper classes. Where girls were married off by their parents before puberty there was very little scope for emotional relationships of the kind depicted in the English novel which the educated Indian enjoyed reading. Romantic love could perhaps be depicted in historical novels in which temporal remoteness helped to minimize the social rigidity and which were not bound by the conventions of realism. In the contemporary Indian setting, however, romantic love could only be illicit, involving either a widow or a courtesan—since only these two categories of women were without legal 'proprietors' and thus seemed to embody a certain amount of unharnessed sexual energy. As both the widow and the courtesan were outside structured society, love of this kind was however doomed from the very beginning. In depicting the man–woman relationship each major Indian writer attempted in his own way to negotiate the demands of the novel with its in-built emphasis on individual self-determination with the intransigence of contemporary social reality.

In order to analyse the different aspects of this problem, four novels, all written in the last quarter of the nineteenth century, are discussed below. These novels are: *Indira* by Bankimchandra Chatterji (Bengali; 1873; revised edn. 1893); *Indulekha* by O. Chandu Menon (Malayalam; 1888); *Pan Lakshyant Kon Gheto* by Harinarayan Apte (Marathi; 1890); and *Umrao Jan Ada* by Mirza Mohammad Hadi Ruswa (Urdu; 1899).[3]

Widely different in theme and technique, these novels have however a few common features: the writers are all major novelists in their respective languages; the protagonists are women (though the writers are all men); each novel deals with the present or with a period within the author's living memory; they are realistic in intention. The last two points become important when it is seen how often prose narratives in the nineteenth century used the conventions of romance and were set in a vague historical past.

The women protagonists are from different regions and diverse social milieus. Indira is the daughter of a wealthy Bengali landlord, Indulekha belongs to a prosperous matrilineal Nair family in Kerala, Yamuna in Apte's novel is a middle class Maharashtrian girl from Pune and Umrao Jan is a courtesan of Lucknow and Kanpur. Three of these novels are narrated in the first person by the protagonists themselves, only *Indulekha* has an omniscient narrator who intervenes occasionally to share an opinion or two with the reader and comment on events. Even in the first person narrations the tone varies considerably. *Indira* is written in the mode of comedy where the heroine's playful high spirits are not suppressed for long even during her misfortunes. The shadow of tragedy falls on Harinarayan Apte's novel *Pan Lakshyant Kon Gheto* from the very beginning. Yamuna's childhood and even the brief happy interlude in Bombay are darkened by an impending disaster. Umrao Jan in Mirza Ruswa's novel has an urbane, cynical attitude towards life, and although she occasionally makes a ritual reference to her 'sin', the tone of her narrative is free from either guilt or self-pity.

Of the four novels only one ends traditionally and happily with a marriage. *Indulekha*, unusually for its time, depicts romantic love in a contemporary situation between two young persons and ends in a marriage that combines individual fulfilment, social sanction and material benefit. Menon could make this realistically feasible by choosing his heroine from the Nair community where, because of the property laws, women have more independence than in most other communities of India. In *Indira* the heroine is already married when the novel begins and her attempts to reclaim an estranged husband constitute the plot. *Pan Lakshyant Kon Gheto* is a bildungsroman tracing the protagonist's growth from childhood to maturity, but marriage is not the climactic event in the heroine's life. Though married at eleven, the meaning of marriage dawns on Yamuna very gradually and widowhood is part of the experience tackled in this dense close-packed narrative of nearly 700 pages. Umrao Jan, by virtue (or vice) of her profession, is outside the structured society where marriages are made, and the novel ends with a sad and cynical acceptance of loneliness.

All four novels are intended to be realistic and each one uses a different device to ensure credibility. In *Indira* and *Pan Lakshyant* the language and style of the first person narrative is meant to constitute subjectivity. Indira's irrepressible vivacity is supported by her rich image-making propensity while her limited feminine horizon is defined by her general acceptance of her narrow social role despite occasional doubts. Yamuna's style in *Pan Lakshyant* lacks sophistication and her ineptitude in verbalizing complex states of mind is only to be expected in a woman without formal education. The prolixity of the novel and its unwieldy frame are also aspects of the characterization of the narrator. The authors of the other two books, Ruswa and Menon, have written at length about realism (see Appendix II) justifying their ways of writing. Although Ruswa did not use the word 'realism' or its Urdu equivalent, he was propagating the need for psychological realism in fiction when he insisted that common events in ordinary lives are full of fictional potential if one has the ability to see them from within. Without mentioning names he makes a dig at contemporaries like Nazir Ahmad who had set their novels in a vague past and not taken any trouble over historical authenticity. With mock humility Ruswa wrote—'Great ability and much labour is required to write a historical novel, and I have neither the ability nor the leisure to do it.' Thus Ruswa's emphasis is on the present time and ordinary life, the two axes on which formal realism of the western variety rests.

Menon states his intentions more clearly than the others in the elaborate preface to *Indulekha* where he extols the superiority of the realistic mode over every other. But there can always be a gap between intention and execution, and the question of how the realistic intentons of these writers relate to their apprehension of reality will be taken up later.

Indira (1873)*

The tone of the first chapter establishes the lively nature of the nineteen-year-old narrator-heroine and her romantic

* See Appendix I for an English translation of the brief first chapter.

yearning to be with her husband. Unlike her proud father she does not understand the value of money, but money and class are important elements throughout the novel. The ostentatious palanquin and the impressive entourage that Indira's husband's new-rich family sends for her invites trouble, and she is abducted by highwaymen while she is on her way to her husband's house. Robbed of her social identity (clothes, jewellery, money), she has to begin life from scratch, entirely on her own. A lawyer's family in Calcutta employs her as a domestic servant and gradually the young mistress becomes her friend. Her employers conspire to bring Indira's husband to the house and thereafter leave the matter to her resources. Indira manages to seduce him without revealing her identity because she knows that no husband will accept a wife who has been taken away by robbers. Once he is completely under her power she takes him into her confidence and, with his help, is accepted by his family. This, in brief, is the outline of the plot.

Since the novel is written in the comic mode (in the revised edition of *Indira*, Bankim quoted Shelley's poem 'Rarely rarely comest thou, Spirit of Delight' to underline the spirit of the work) the author touches only very lightly upon the moral and social dilemmas of the young heroine. The emphasis is on her clever handling of problems rather than on the basic nature of these problems; yet there are a few uneasy moments where the questioning of values cannot be altogether avoided. For example, when Indira has succeeded in snaring her husband she has moments of doubt about the worth of a man who is so easily tempted. While Suhasini, her mistress and friend, decks her up for the secret assignation at night, Indira says: 'I am happy, but I can't think very highly of him. I know there is nothing wrong in what I am doing because he is my husband, but he does not know that I am his wife. How can I respect a man who agrees so easily to have a secret appointment with another man's wife?' Suhasini tries to justify his behaviour by reminding Indira that he is a man, and that he does not have a wife at home to be loyal to. Indira replies, 'I too don't have a husband' and asks a crucial question: 'Is it so difficult for a man to control his desires?' After reaching this uncomfortable point

the novel moves back to the comic level and when Suhasini
suggests that a woman must please a man by looking after the
small details of his comfort, Indira's spirited rejoinder is:

'But that is the work of a slave. Am I trying to win him back only to
show him how good I can be as a slave?'
'What are we but slaves?'
'When there is love between us I will not mind doing a slave's work.
I will press his feet, fan him and prepare his pan leaf. Not until then.'[4]

After winning back her husband Indira is not expected to
question his individual worth. Upendra the husband is to take
precedence over Upendra the man. What emerges of his character is not very positive. Indira notices his naive gullibility in
contrast to the sophistication of the English-educated men she
has seen in her employer's house in Calcutta, but she does not
allow herself to be critical. She knows that she must love her
husband irrespective of his personal worth.

There is a recognizable gap between the romantic yearning
of the girl who was eager to meet her unknown husband
(Chapters 1 and 2) and the woman who settles down to live
with her reclaimed husband at the end of the novel. On the
surface the novel ends happily enough and the darkening of
the tone between the beginning and the end is scarcely
admitted, but the quality of Indira's language and the imagery
she uses suggest a sense of loss. When she makes her first
journey she is a girl brimming over with the joy of youth, waiting
for the man who will make her life complete. At the end
of the novel when she makes the same journey with her
husband she seems more subdued than fulfilled. She herself
tries to explain the differences by contrasting the joy of
imagination with the happiness of acquisition. Her first journey
was like poetry, the other like wealth. 'Can a rich man's wealth
be ever compared to poetry? . . . Can the realization of a dream
be as delightful as the dream itself?'[5] In her characteristically
exuberant and metaphor-laden manner Indira goes on to
speculate on the similarity between money and the colour of
the sky. The sky is not blue, she says, 'we only think it is so'. 'So
with money. Money is not happiness, we only think it is so.
Poetry is happiness because it generates hope. Money can only

be consumed, and not everyone knows how to consume it either. So many rich men spend their lives protecting money like treasury guards.'[6]

The imagery given to Indira at this point reveals more than the overt intent of the plot. The development of the plot can be seen in four phases: anticipation of happiness, frustration and loss of happiness, attempts to gain happiness, achievement of happiness. The word 'happiness' here can easily be substituted by the word 'husband' because in the conscious social norms of Bankim's world there can be no doubt about their equation. Achievement ought to be better than anticipation, and if the narrative is seen as a graph, the novel should end at a higher point of bliss than where it begins. Yet in its actual structure, realized through language, imagery and tone, the achieved happiness of the end falls far below the anticipation. One does not know if Bankim was conscious of this effect, or if it can be explained away merely by saying that adult reality seldom matches adolescent dream. Whatever the author's intentions, the theme of the novel can be seen as a transformation of the romantic longing of an unfettered individual into the complacent purr of a tamed wife who has found her gilded cage.

Yet in the last paragraph of the novel, which by the logic of the surface plot ought to have been the happiest, Indira's smugness seems to be undermined by a vague nostalgia: 'I have not forgotten Suhasini. I will never forget her as long as I am alive.'[7] Suhasini was Indira's friend (initially her employer) during her days of hardship. This was one relationship in Indira's life where the ties were not familial or social. The two girls related to each other entirely as two individuals. Suhasini had brought Indira home knowing nothing about her. Indira, at the time, wore only a coarse sari and had no social identity markers, such as jewellery. By the time Suhasini discovered marks on Indira's skin that indicated a habitual wearing of jewellery, revealing her rich parentage, they were already friends. No other relationship in the novel has the same intensity as that between Indira and Suhasini. In describing Suhasini's beauty Indira becomes lyrical, with images of the lotus, the waves of a river and the swaying branches of a mango tree. Today readers might see traces of lesbianism in this attachment, and it is well-known that in a gender-segregated situation, friendship between persons of the same sex tends to gain in intensity. Further, it must be remembered that even though the narrative voice is a woman's, it merely projects the male author's view of a woman's sensibility. Here, as elsewhere in the novel, in the description of female beauty the author seems to take over from the narrator.

Like many of Bankim's other novels (*Kapalkundala, Visha Vriksha, Durgeshnandini*) *Indira* begins with a journey. A journey is a specially significant experience for a woman who normally inhabits an enclosed space. The mobility and the freedom of the road liberate her for a short period from the inhibitive social structure. ('I wish I could be a bird.') There are three journeys in the novel. After being abducted and robbed by the highwaymen Indira is taken to Calcutta on a boat by a charitable family. During this second journey she is enthralled by the beauty of the Ganga and looks at everything with wonder and delight despite her present misery. She is amused by the flirtatious songs of two little girls on the river bank. The girls sing of how they will shake their ankle-bells and strut brazenly. Indira's prosaic companion is annoyed at the vulgarity of the song, but the song makes Indira wonder about the traditional definition of decency. The author makes the ankle-bell-shaking song part of the design of the novel, prefiguring Indira's own later action. The third journey is undertaken when Indira's husband brings her home at the end. 'Travelling with my husband was a happy occasion no doubt, but my first journey was joyous in a different way.'[8] Thus a good part of the pattern of the novel is traced through the motifs of journey.

A long-distance journey in the nineteenth century was an adventure fraught with hazards, and it is no coincidence that all four novels under discussion use robberies as part of the plot. Except in *Indulekha* where the event takes place in a railway station, the other novels feature highway robberies. This was a common enough danger at the time, and in Bankim's novels this situation is often used as a plot device. In *Kapalkundala* Motibibi meets Nabakumar because of a robbery; in *Debi Chaudhurani* and *Anandamath* the central characters are members of bandit gangs who plunder for altruistic purposes. Unlike *Indira*, which is set in a period close to its own times, these three novels of Bankim are set in the past: particularly the latter two which are set in a time when lawlessness had reached its peak in Bengal. The British had attempted to impose law and order, but even in the mid-nineteenth century when Indira's husband left for the 'far away Punjab' we are told 'the road going west was dangerous and difficult'.[9] Highway robbery was a common occurrence till late in the nineteenth century and memories of *thuggees* and *bargis* (marauders who travelled across the country) have remained even to this day part of the legends and nursery rhymes of Bengal.

Apart from being a realistic reflection of life, robbery also served as a fictional device by unsettling the status quo and, as in *Indira*, by setting the action in motion. As in the game of snakes and ladders, from a very high position the heroine falls down to reach rock bottom, from whence she has to slowly make her way up, depending entirely on her own resources. This then was a strategy of divesting the protagonist of the protective social crust and seeing her as an individual, un-supported by the props of family, class and caste.

Indulekha (1888)

If in *Indira* there is a subtle tension between the author's overt social attitudes (upholding of order and orthodoxy) and the romantic and individualistic undercurrents revealed by tone and imagery, in *Indulekha* the conflict is between the conscious aesthetic intentions of the writer derived from his reading of western literature on the one hand and the gravitational pull of traditional narrative conventions on the other.

Indulekha began as an adaptation of Disraeli's *Henrietta Temple*, a minor and forgotten Victorian novel, though in the completed work not much trace remains of the model. It is a strange irony of the colonial situation that some of India's major nineteenth-century writers looked up to second or third-rate British writers as models and often ended up achieving much more than those they emulated. Apte modelled one of his novels on W.M. Reynold's *The Seamstress*, a Victorian bestseller, and Bankim acknowledged his debt to Bulwer Lytton and Wilkie Collins in the preface to *Rajani*. Menon far surpassed his modest aim of adapting Disraeli and ended by writing the first major novel in Malayalam.

Indulekha was written with two specific objectives: to introduce a new literary form in Malayalam—a realistic narrative of contemporary life—and to show the necessity and advantages of English education, especially for women. The reasons for the first objective are explained at length in the Introduction (see Appendix II), where Menon berates the mythic imagination of the Hindus, extols the western mode of realism in painting and extends the same argument to literature. Modes in literature and painting have often echoed each other. In the eighteenth century Fielding and Hogarth referred to each other to justify their satiric intentions and emphasize their views of reality. The critical writings of Henry James and Virginia Woolf are similarly full of comparisons between fiction and painting. Literature and painting often reflect the same ethos and thus impressionism, surrealism and cubism can all be shown to have parallels in literature. Menon rejects the traditional Indian modes of both painting and story-telling to adopt what he considers a superior mode. Yet his actual work often veers towards the pre-novel forms of story-telling and away from the realistic technique that he so admires.

Menon's young lovers—Indulekha and Madhavan—are both paragons of beauty and virtue, and their courtship is presented in a stylized manner (her heart is flint, he cannot sleep at night, they express their love in poetry). There is no characterization in the realist sense and the delay in their union is caused purely by external circumstance. There is no conflict within the individual or between the individual and society, the adversaries often being either illusory or comic figures. The action is propelled forward by accidents and coincidences. Once separated, the heroine languishes in illness while the hero goes on his picaresque journey, following predictable fairy and folk tale conventions.

But the ironic gap between the author's intention and his execution is never more clear than at the end, where Menon writes:

Now my story is ended ... All the characters mentioned in this book are still alive. Madhavan has now attained a high rank in the Civil Service and he and Indulekha are blessed with two children, one a daughter and the other a son, both beautiful as the harvest moon. Madhavan, by his industry, uprightness and ability, and Indulekha by her devotion to her children and her husband, have reached the summit of human happiness, and may God bless us and all who read this tale.[10]

The last line in this paragraph is very similar to the ending of the oral recitals of the puranas where the telling of and the listening to a story are ritual acts. The teller and the audience all acquire merit through the performance.[11] Even if the recurrent authorial intrusions in *Indulekha* are not seen as part of the oral narrative tradition ('Now my story grows sad,' 'Now to return to our story,' 'Now my story ends'—Victorian novels too abound in such interventions) the *katha*-like ending of the novel links it to the tradition of sacred recitals, quite different from the secular realistic tradition of the western novel. In the fairy-tale-like ending the hero and heroine achieve the 'summit of happiness' and live happily thereafter. What could be a greater happiness in late nineteenth-century India than getting into government service?

The second objective of the writer is also stated in the Introduction:

as one of my objects in writing this book is to illustrate how a young Malayalee woman, possessing, in addition to her natural personal charms and intellectual culture, a knowledge of the English language would conduct herself in matters of supreme interest to her, such as choosing of a partner in life, I have thought it necessary that my Indulekha should be conversant with the richest language of the world.[12]

The learning of English is not just the acquisition of a linguistic skill, it is the imbibing of a new system of values and gaining self-confidence. Demure and respectful to her elders, Indulekha can still challenge the code whereby marriage is an economic transaction between families and not just a matter of individual choice. Her grandmother explains the traditional obligation of a Nair woman:

When a woman is beautiful and clever she must do some good to her family. She ought to make a good match. Money, my dear, money is the great thing ... The girls in our family have always been attractive, but none of them has ever yet been so attractive as you my child ...[13]

Without obviously disobeying her Indulekha quietly manages to ward off through ready wit and a sense of humour the rich Nambudiri suitor sent to court her. She is in love with her cousin, but marries him only when this act is sanctioned by the community – a sanction that she manages to get not through rebellion but quiet self-assertion. Even Kesavan Nambudiri who has been opposed to women learning English has to grudgingly admit the power of education:

I am convinced now that women who have learnt English are beyond our comprehension altogether. There's Panchu Menon, who isn't afraid of any one in the wide world, but he shivers and shakes before this chit of a granddaughter of his ... I begin to think now that perhaps I must have been mistaken.[14]

Throughout the novel English education is given an almost sacred value. A part of the power of the British in India was technological superiority. This has always been the basis of colonial domination (one recalls the magical power of gun-powder in *Robinson Crusoe*, an early paradigm of colonial experience) and in *Indulekha* Kesavan Nambudiri is awed by the thread factory in Calicut where the wheels move 'as if they heard the word of command'. He speculates on the power of the smoke that comes out of the factory and the dark sacrifices that must be made to generate this power. Kesavan Nambudiri's confusion is between two systems—the rational and the magical, or the natural and the supernatural (factory smoke—sacrificial fire). His wife knows that English education is the only means of demystification. Indulekha is not awed by anything because she can even explain 'the principles on which the railway is driven.'

Indulekha is not a realistic character by the author's own admission: '... my object is to write a novel after the English fashion, and it is evident that no Malayalee lady can fill the role of the heroine of such a story. My Indulekha is not, therefore, an ordinary Malayalee lady.'[15] She is an illustration of what the Nair woman can become. Thus she belongs more to the tradition of 'exemplum' literature (for example, Chaucer's 'Knight's Tale'), where the ideal is more important than the actual, than to the realistic novel.

The position of women in society seems to have been a subject of considerable concern to the nineteenth-century novelist in India, as also to social reformers and pamphleteers. This was a part of the new awareness of human rights and of the concept of individuals as autonomous units. Women's education and the amelioration of the plight of widows were two major aspects of the social reform movement. In Bengal Ishwarchandra Vidyasagar agitated for the remarriage of child-widows and his treatises (*Shakuntala* and *Seetar Banabas*) were meant to arouse consciousness of the 'woman' question. In the Telugu journal *Tatva Bodhini* (1864–70) problems of widow remarriage were discussed. In Maharashtra this was already a controversial issue in the 1850s when Baba Padmanji wrote *Yamuna Paryatan*, a narrative solely concerned with the plight of widows. In Urdu Nazir Ahmad wrote his first book *Mirat-ul-Arus* (1869),[16] propagating the need for women to be educated, though of course he did not mention English as a necessary skill.

In *Indulekha* the milieu is a matrilineal community and consequently the women are less subjugated than in a patriarchy. But even here one finds prospective husbands like Suri Nambudiripad coming to buy a bride, and when Indulekha seems inaccessible his roving eyes fall upon her mother. Finally a little girl is given to him as a bride 'just as she had been a kitten about the house'. Menon exhorts at this point:

My beloved countrywomen, are you not ashamed of this? Some of you have studied Sanskrit; and some music, but these attainments are not enough. If you wish to really enlighten your minds, you must learn English, whereby alone you can learn many things which you ought to know in these days and by such knowledge alone can you grasp the truth that you are of the same creation as men, that you are free agents as men, that women are not the slaves of men. (p. 369)

The professed realistic intentions of the Introduction are left far behind. Menon ends up using his novel for conveying a moral, as in the familiar indigenous narrative tradition.

Although the novel presents a detailed account of one Nair family, a larger picture of late nineteenth-century Kerala life emerges through the interaction between the upper castes (specially Nairs and Nambudiris) and between the Indians and the British. The landed gentry shows little initiative in developing production. For paltry payments Suri Nambudiripad signs away large areas to European planters. The British are the greatest bestowers of patronage and status and Madhavan's success is due to the timely recognition of his merit by a sympathetic British officer. When Nambudiripad hears that his opponent in a court case has appointed a European barrister his immediate reaction is: 'Then we must have a European too'. In the long digressive eighteenth chapter of the novel (constituting fifty-four pages) two subjects of great interest to the late nineteenth-century Indian—religion and politics—are thrashed out in an elaborate argument among Madhavan, his father and a friend. Even though (as the author himself apologetically admits) the chapter does not have much bearing on the plot of the novel, it captures the cultural and intellectual ambience and places the novel firmly in its time—a period of active reform movements in Indian society and the emergence of the Indian National Congress.

Another digression in the novel is Madhavan's long journey, including his stay in Calcutta. This, however, was necessary from the point of view of the plot because the hero and the heroine had to be separated. Since Indulekha's situation was static she could be left alone while the reader followed the peripatetic hero who, during the course of his adventures, emerges as an individual outside the social structure that has so far defined his role. In the voyage from Bombay to Calcutta, when his ship passed the Malabar coast, 'the yearning which overcame Madhavan at the sight of his native land was unutterable'. His knowledge (mistaken as it turned out later) that Indulekha was unattainable, and his sense of adventure, made him continue the journey.

Like other elements in the novel this fictional journey also has a British model. Mr Gilham, Madhavan's superior officer, suggests to him that he should travel for a while because 'it is a very usual thing for young men in England to make a tour abroad after leaving the university before they take up a profession' (p. 260). Since Madhavan was advised against going to Europe in winter he began his grand tour of India—the northern part of which was almost as alien to a man from Malabar in those days as England. We are told that during this journey Madhavan wore boots and clothes 'as worn by English men'.

Madhavan's journey seems interesting to the present-day reader for two different reasons. First, in the nineteenth century the non-historical novel rarely moved beyond the confines of the linguistic areas to which it belonged. Reading Bankim's or Apte's novels one seldom gets the impression that India is a varied country in terms of language and culture, and the vast space of the subcontinent was never used for fictional purposes. *Indulekha* is unusual in moving out of Kerala, first to Bombay, then to Calcutta and other parts of northern India, and in using characters who are not from Malabar. Secondly, since Menon was forging a new form of story-telling in Malayalam, it was also his concern to see that the reader was enticed to stay with the story. Thus the conducted tour of the country to which the reader is taken along with Madhavan serves the same purpose as location shooting in Europe in Indian commercial films today—it provides a vicarious experience to the untravelled audience. Hence the Bombay harbour is described minutely with appropriate awe and the riches of Calcutta are presented in all their splendour. The description of the palace of Govind Sen and his brother Chittaprasad Sen in Calcutta extends over three pages. Vivid details of both the interior and the exterior—'broad carriage drive made of the finest gravel and rolled smooth, swept in a graceful curve up to the hall door ... fountains of marble cunningly wrought ... sofas, covered with rich velvet and brocade ... marble topped tables and tables fashioned in the English style out of ivory and out of wood ... numberless huge

mirrors set in golden frames...' (p. 273)—testify to the legendary wealth and luxury that rich merchants in Calcutta enjoyed
in the nineteenth century. This bears comparison to a Calcutta
interior described by Bankim around the same time which,
though not so resplendent, appears equally hybrid: 'Chairs,
tables...American clocks, glassware in variegated hues,
pictures for which the *Illustrated London News* is liberally laid
under contribution...bookshelves filled with Reynold's
Mysteries, Tom Paine's *Age of Reason* and the complete
Poetical Works of Lord Byron, English musical-boxes, compose
the fashionable furniture of the young Bengal.' Madhavan's
hosts, however, belong not to the fashionable 'Young Bengal'—
another name for a westernized rebellious group, but to the true
aristrocracy, being "among the leading millionaires of Calcutta.'
Coming from agrarian Malabar Madhavan is dazzled by the
wealth of the colonial city of Calcutta, and the author intervenes
to make a distinction between the rich men in Madras and in
Calcutta: 'In Madras the lord of five lakhs or half a million rupees
is considered a worthy of the first magnitude, while in Bengal
where the first class is composed of the owners of five crores or
fifty million rupees, he would be relegated to the fourth class' —
thus making it clear that the wealth of Suri Nambudiripad is of
quite a different order from the Sens of Calcutta.

On his way back from Calcutta Madhavan is robbed of all his
belongings by an imposter who pretends to be the sub-judge
of Allahabad. Menon achieves realism graphically in the
description of this incident, down to the details of the food in
the railway refreshment room and the apathy of the railway
police. The police take no interest in Madhavan's loss until
they discover that he has contacts with a rich business family, a
familiar enough situation even today. But as a government
servant himself Menon felt the need to add quickly: 'Of course
the place where the theft was committed was beyond the limits
of British India, and the police in question were not our British
police.' (p. 285)

Menon's overt aim of writing a realistic novel 'in the English
manner' was occasionally overshadowed by the influence of
traditional narrative modes which depended on non-mimetic

devices like coincidence and prophetic dreams. Menon's contemporary Bankim also used dreams as a fictional device (for example, in *Visha Vriksha*), but Menon was anxious to justify his use of dreams 'to reveal the present of the past, or to presage the future' (p. 360) since he knew that this was in direct opposition to the rationalism which underlay the realistic model he intended to emulate. He wrote in a digression: 'Frequently our dreams are altogether unreal, but on the other hand it is recorded how an Englishman, tired after a day's shooting, fell asleep in his tent and dreamt that he was attacked by a cobra, with hood raised to strike. So vivid was the dream that the man thought he was actually bitten and woke terrified, when he saw a snake gliding along within a few feet of his camp cot'. (p. 361). The important point here is that the incident happened to an Englishman, hence its veracity cannot be doubted.

Menon's uncritical admiration of anything British permeates his novel at many levels. In order to forestall any misunderstanding that his caricature of a Nambudiri brahman might create among that community he cites again the superior example of the English:

In English novels the characters, male and female are all drawn from various ranks of European society and in some books even living celebrities are occasionally made the subject of censure, ridicule or praise, but unless the story is prompted by malice, no one thinks of taking offence at, or quarreling with, the representations given therein. (p. 95)

An awareness of the gap between the British ideal and the imperfect Indian reality runs through Menon's novel. He loses no opportunity to extol everything English—from traits of character to literary fashions—yet his unconscious allegiance to traditional Indian modes of narration often comes to the surface. Thus in his case the rhetoric of prior intent tends to contradict his actual fictional achievement.

Pan Lakshyant Kon Gheto (1890)
There is a similarity in the narrative frameworks of *Pan Lakshyant Kon Gheto* and *Umrao Jan Ada*, although they are quite different from each other in subject matter and tone. Apte,

in the preface to *Pan Lakshyant,* tells the reader how he came across a manuscript written by an unfortunate girl called Yamuna and edited it for publication. Ruswa, in the preface to his Urdu novel, tells the reader how he met Umrao when she was leading a quiet retired life and how he persuaded the courtesan to tell him her life story, which he subsequently wrote down. In the early stages of the English novel Defoe had used a similar strategy in *Moll Flanders* when, in the preface, he told the reader that he was merely editing the memoirs of a woman criminal whom he met at Newgate prison.

The autobiographical technique is in each case intended to create an effect of verisimilitude, establishing both the narrator's character and the solidity of her social milieu. In Apte's novel absolute authenticity of the narrator's voice is more important than formal neatness. Yamuna is a middle-class Pune brahman girl and her style of narration has the absence of sophistication that can be expected in a woman without formal education. Her lack of adequate verbal resources is indicated through the repetition of certain phrases: 'I don't know how to say this', 'my words fail me', 'I cannot describe'. Although she tries to maintain a chronological order she often mixes up sequences unwittingly and then admonishes herself for her ineptitude. There are two kinds of 'now' in her story—the time of writing, and the time when the events actually take place. But this amateurishness is part of the author's conscious design and we find Yamuna fairly successful in modulating her point of view from that of a child to that of a girl, and later to that of a woman whose consciousness expands to discover new horizons. The child Yamuna's point of view is limited—she does not understand the problems that beset her mother and is mystified at the sudden moves of the family; but an adult Yamuna can confidently comment on social injustice or discuss with the reader her narrative problems. While describing the two years of happiness in her life—the time she spent with her husband in Bombay—she admits that conveying the experience of happiness is far more difficult than conveying that of adversity because happiness is much more a state of mind than an activity.

Yamuna grows up in the enclosed space of a joint family in Pune. After her marriage she goes into an even more claustrophobic world where nothing is private, where conversations are overheard, where there is constant intrigue and politicking within the family, and where a new bride is an object for everybody to exploit. She does not know how narrow this life is until much later when her vision widens through books and contact with the outside world. As a contrast to the circumscribed worlds of Sadashivpeth or Tulshibag in Pune, there is freedom in Bombay. Even before Yamuna actually reaches Bombay, the city has been part of her day dreams for some years. She had known that her husband would take her there after he finished his studies and that she would emerge out of the octopus-like clutches of a joint family to breathe and to set up her own household. When she finally arrives in Bombay she is overwhelmed by its size and grandeur: 'I used to think our Poona was a big city. But this was not a city—it was like a different country....I sat down foolishly at the Victoria Terminus and looked around with awe and wonder'.[17] This is comparable to Indira arriving in Calcutta and being similarly overwhelmed. In both cases the big city gives the women opportunities to be their own selves. While Yamuna learns to read, write and think, to discuss things with others and talk about the condition of society, Indira, some twenty years earlier, used her own initiative to reclaim her social position. The movements are in opposite directions—in one Yamuna moves out of a closed world towards an open one and in the other Indira chooses to return to the security of a closed world. Although Bombay is the world of awakening and individualism, for Yamuna orthodox Pune is never far away. Every time she returns to Pune for a vacation she is sucked back into the turgid pool of narrow views, double standards and the tyranny of public opinion. The tragedy of Yamuna's life is that once having breathed pure air she has to return to a stagnant society, and after her liberal husband's death becomes a sacrificial victim of the old oppressive ways.

If the word feminism had been current in the last decade of the nineteenth century, *Pan Lakshyant* might have been called

a feminist novel. At first docile and limited like all other women
of her social class, Yamuna gradually grows into a new
awareness and begins to ask the kind of radical questions that
none of Bankim's heroines are permitted to ask. In all of
Bankim's thirteen novels (except *Kapalkundala* where the
heroine asks, 'Is marriage a prison?') women accept society's
valuation of them. The mature Yamuna, on the other hand,
recalling the bride-showing incidents of her childhood, cannot
suppress her indignation. She recalls how at the age of eleven
she was overjoyed when the groom's family came to see her
because it was an occasion to dress up and be part of the adult
world. The mature Yamuna sees the same ritual in terms of
the purchase of a domestic animal. Once the child-bride is taken
to her hunband's house she is like a puppy who is at first
fed on milk and rice. 'At least the puppy is loved. We cannot
always be sure of that. We learn to survive by fawning on the
master with a show of devotion and flattery'. Such moments of
brutal honesty are usually followed by an apology in the novel.
The comparison with the puppy concludes with this remark:

Every intelligent reader can see that the comparison can be stretched
quite far. But those who will not see the point will be annoyed with me.
Hence I must let go of this simile and proceed with my story.[18]

Though written in a densely textured realistic manner, the
ancestry of *Pan Lakshyant* can be traced back to the rudimentary first novel in Marathi written thirty-three years earlier.
More like a tract than a novel *Yamuna Paryatan* (1857) also
dealt with the exploitation of women, and particularly that of
widows in Hindu society. Apart from featuring heroines who
have the same name, the two works have other incidental similarities: both the Yamunas are married to liberal husbands, both
eventually become widows and in both cases the central
characters are sufferers as well as witnesses to the suffering of
other women. But *Pan Lakshyant* is much more than a 'thesis'
novel; it captures in a vast canvas the subtle enmeshing of
the individual's life with society, evoking in the process the
two major forces of nineteenth-century Maharashtra: brahman
orthodoxy and reform movements. Most movements of social
change in India have emphasized the need for women's

education and, as in *Indulekha*, in *Pan Lakshyant* English education becomes closely associated with the process of achieving a new social and individual identity: thus after learning to read and write Marathi, Yamuna begins lessons in English, though her studies are cut short by the death of Raghupat Rao, her husband and mentor.

Apte manages to give Yamuna's relationship with her husband a personalized quality. Yamuna loves and respects Raghupat Rao as a person, unlike Indira who is interested in the abstract idea of a husband. Marriage is important in *Pan Lakshyant* as a social institution (the novel begins with a doll's wedding which reproduces in a small scale the bickerings and politics of real weddings) and, except in the case of Yamuna and Raghupat Rao and two of their friends in Bombay, hardly ever results in emotional fulfilment on a personal level. Physical relationships between a man and a woman are seldom referred to except as something to be afraid of, as when Durgi's husband tries to assault his child-bride. The middle-aged widower Shankar Thakur marries a young girl and is inordinately eager to have the 'blessing of the womb' ceremony performed soon — an euphemism for the ritual which licenses him to cohabit with his wife. The horror of such bestial marriages is built up very gradually. The first shock comes when Yamuna's father, only a few weeks after her mother's death, brings home a wife hardly older than Yamuna. As Yamuna grows in her experience of life she begins to see the disparity between the situation of a woman who loses her husband and of a man who loses his wife, and that a double standard in this regard is universally accepted. A widow's life is a long series of privations, and the shaving of her head is a symbolic rite to desex her. A man, on the other hand, whatever his age, is encouraged to marry even before his dead wife's obsequies are over. The ironic force of this is intensified in the climactic action of the novel. Yamuna refuses to have her hair shaved after her husband's death, creating a public scandal. No priest can come to bless Shankar Thakur's bride's womb so long as an unshaven widow lives in the house. So Yamuna is locked up in a room and her head forcibly shaved to enable a man three times her age to gratify his desires. The manner in which Yamuna's hair is shorn is reminiscent of an animal sacrifice: first she becomes unconscious in terror and then her oppressors take advantage of her unconsciousness.

The head-shaving incident is never named in the novel but always referred to as the 'cruel deed' or 'the horrible act'. The people within the social milieu in which this Marathi novel was read knew the dreadful associations of this event all too well. Therefore when Yamuna writes the account of her life in the last days of her ailment she avoids naming or describing this event directly, taking recourse to suggestions and innuendoes.

Yamuna is persuaded by her brother to write the story of her life so that others may know of the injustices perpetrated in her society. She agrees to do this although she is more cynical than her brother about the objectives:

Today my brother said that if the women who spend their lives in such misery learn to read and write, or if they ask someone who can read or write to write out their experiences of life, then these accounts can help in ameliorating the condition of women. My brother is mistaken in this. Can't people see for themselves how we poor helpless women live? ... Can't they see how we are exploited by the self righteous guardians of morality and religion? If one opens his eyes, he can see. If one wants to find out, he can. But who wants to see? Whoever bothers to notice anything?[19]

The last sentence of this passage gives the novel its title. Though apparently loose and meandering the novel does have a conscious design. Around Yamuna's story several situations are presented that counterpoint the central plot. Yamuna's husband widens her awareness of life, but her childhood friend Durgi's the world is gradually stifled by her worthless husband. While Yamuna blossoms forth in the open atmosphere of Bombay and overcomes her inhibitions, her brother's wife in a similar situation withdraws into a tighter orthodoxy. When a grown up Yamuna returns briefly to her grandfather's village she observes: 'There has been so much reform all over the country. There have been railroads, etc. But the road to our village remains exactly as it was. There is no change here.' The stasis is not confined only to the road. When Yamuna tries to locate her childhood playmates in the village she finds that one is a widow who makes a living by cooking for her brother-in-law's family, another is burdened with repeated motherhood at an early age and a third has died at childbirth.

Untimely death stalks the entire novel. This can be seen as a crude fictional strategy to intensify the tragic effect but it may just as well be a realistic presentation of existing conditions at a time when epidemics and childbirths killed many, and certain ailments could not be cured. Both Yamuna and her husband Raghupat Rao died young, and both of tuberculosis—which continued to be the cause of premature death well into the age of Saratchandra.

Unlike the author of *Indulekha*, the author of *Pan Lakshyant* professes neither realism nor any social purpose, but by choosing the right narrative voice he effortlessly achieves both. Apte does not put forward any fictional theory but succeeds in creating a convincing character through whose perceptions the complex interactions between an individual and the larger forces of history become vivid.

Umrao Jan Ada (1899)

The waves of social reform and women's awakening that were washing the coastal areas of India (Bengal, Maharashtra and Kerala) left the inland region untouched. Feudal Lucknow, where Ruswa's novel is set, was a different world in which values of an earlier age continued into a decadent milieu of nawabs and *tawaifs* in a glittering but fragile texture of music and poetry. Yet the ripples of the larger historical transformation were not totally invisible. *Umrao Jan Ada* describes a period when the monarchy of the Nawab of Awadh was giving way to British rule. The revolt of 1857 comes about halfway through the novel, when Umrao Jan is at the height of her power. Before 1857 the capital city of Delhi had declined and Lucknow was so much the centre of prosperity, culture and refinement that even minor poets from small towns pretended to belong to Lucknow. 'Many make their living on the name of Lucknow as I did when I was in Kanpur', says Umrao Jan. She becomes famous in Kanpur and in Faizabad as the courtesan of Lucknow. By the end of her career, about the time she narrates her life story to Ruswa, Lucknow too had begun to decline and the seat of prosperity had shifted to the Deccan. The lawlessness in the countryside that was common in much of North India in pre-British days is reflected in this novel too, although towards the end Umrao acknowledges the welcome changes brought about by the benign British rulers, making life more orderly in Lucknow: 'Wide open roads were being laid out in many places and the by-lanes were being paved with hard bricks. Drains and gutters were swept regularly. Lucknow had become a different place'. (p. 173) The presence of the British may be seen as the immediate or remote backdrop to all four novels discussed, but other than this the novels do not reveal any awareness of the multiplicity of class, caste or race within Indian society. Each community or caste seems to live its own autonomous life with seldom any outside contact—other than what is socially approved. Reading these novels one would think that in the Pune or Bombay of Harinarain Apte's time no other caste but brahmans existed: Indira belongs to a wealthy kayastha family of Bengal and she finds employment in another kayastha family until she recovers her husband. In *Indulekha* we find some interaction between castes, but marriages between the Nairs and the Nambudiris were sanctioned by society. Given the historical period the encounter between the Keralan Madhavan and the Bengali family of Calcutta is more interesting. *Umrao Jan Ada* hardly mentions a character who is not Muslim (except Ramdei, a girl abducted along with Umrao). Among the Muslims there are the rich and the poor but the social and ethical code that they live by is not very different. Only in the Faiz Ali episode does one see a fleeting glimpse of a conflict of class values. Faiz Ali has money but he lives by robbery and is unused to the niceties of an elegant establishment like Khanum's. Altogether gauche in these rarefied surroundings, he talks crudely about payments for 'services rendered'. Umrao's decision to run away with him is evidently a wild error, based not on love but on a delusion produced by the reading of romances:

I felt very sorry for him. When I used to read in old legends of how some sweethearts betrayed their lovers, I used to feel distressed and cursed them. I felt that if I did not stand by Faiz Ali I would be doing much the same. (p. 122)

Class and place snobbery figure marginally in another place, when the small town rich man Rakhan Mian is described some-what along the lines of country boobies in Restoration comedy. Later, Ruswa and Umrao discuss how parvenus who come from small towns into the fashionable world tend to be more deadly than their city counterpart. But the differences remain at a superficial or comic level—the basic values are the same.

This homogeneity in the fictional world was not very con-ducive to the development of the novel as a genre. In the West the genre had, in its early stages, thrived on the clash between moral and social values of different classes. It is a question worth investigating whether the homogeneity depicted in Indian fiction was a realistic representation of Indian society, where people lived in compartments, or whether it was a convenient device of make belief—a strategy to leave out inconvenient complexities in order to simplify the narrative.

Indian novels of the nineteenth century often invite compari-son with English novels of the eighteenth century in so far as in both cases the genre was in its formative stage. It has been shown that in Richardson's *Pamela* the struggle is not only between two individuals (Pamela and her employer Mr B) but 'between two opposed conceptions of sex and marriage held by two different social classes, and between two concep-tions of the masculine and feminine roles'.[20] Defoe's first two novels, *Robinson Crusoe* and *Moll Flanders*, were triumphant assertions of economic individualism, and they were specially important because the individual in one case happened to belong to the middle class which had recently achieved greater political and economic power, and in the other to the criminal class, outside structured society, for whom economic survival was a higher value than morality. The novels thus reflected the social transformations that were to accom-pany the Industrial Revolution and colonial expansion.

It is the novelist's special prerogative to be able to see the individual and the social process as part of a dynamic continuum. Apte achieves some of this through the character of

Yamuna, who is the product as well as the victim of concrete
social circumstances but who becomes a recognizable individual through her authentic narrative voice and the delineation
of the minutely recorded specific environment which shaped
her.

Umrao Jan also becomes a believable character through her
unique narrative voice. The whole story is recollected in tran-
quility (or cynical resignation) in her middle years when she
is reflective and mature, yet she manages to give each part of
the story an appropriate point of view. Before she was
kidnapped she was only a child—and all her memories of those
days refer to the external surface of life, to food, clothes, to the
shape, colour and texture of material objects:

I wore tight-fitting red silk pyjamas with a waist band of twill. My
blouse was made of nainsook and my dupatta of fine muslin. I wore
three silver bangles on each arm, a gold necklace around my neck,
and a gold ring in the nose (other girls wore silver nose rings). My ears
had just been pierced and had blue threads strung through the lobes.
An order for gold ear-rings had been placed with the goldsmith.[21]

She was betrothed to her cousin at the time of her abduction
and her girlish excitement about marriage is similar to little
Yamuna's enthusiasm in Apte's novel: for both it is an occasion
to dress up and be made much of. The girl-bride's point of view
is further brought out in her sense of childish triumph that her
fiance is fair while her playmate Kareeman's fiance is dark.

Umrao Jan is never shown to be repulsed by the kind of life
she is forced to lead. The philosophic reflections about sin
and god are afterthoughts that belong to her middle age. In her
youth she gets into the spirit of the courtesan's world with
eagerness and gusto. Until she is old enough to entertain 'guests'
of her own she smarts with envy and desire, looking at the older
girls dallying with their clients in resplendent clothes: 'I am
ashamed to admit it, but the truth is that I wanted all these girls'
lovers to love and be willing to lay down their lives for me only
... But alas no one bothered about me'.[22] Ruswa's version of
Umrao's life is narrated in fourteen chapters; the fifteenth is
reserved for Umrao's own comment on Ruswa's account and
her general observations on life. The earlier chapters are

occasionally punctuated by Umrao's witty repartee with Ruswa and never show any maudlin self pity. Though she insists again and again that she was driven to becoming a courtesan by circumstances, she also seems to suggest that like people in other walks of life a courtesan lives by a code of conduct and that she has been true to her professional ethics. When she calls herself a sinner she merely bows to conventions of rhetoric: her matter-of-fact tone of narration seldom betrays any serious feeling of guilt.

It is not clear whether young Ameeran, before she became Umrao Jan, resented being sold like cattle, but she confesses that it did not take her very long to begin to enjoy her new life. As a mature woman she recollects the transaction vividly. Khanum had asked—

'Husaini, this girl does not seem too dear for the price we have paid.'

'Dear? I wouldn't say you got her very cheap either.'

While Umrao Jan was bought for 125 rupees, in later years she herself bought a girl for only one rupee (during a famine). There is no direct condemnation implied either by Umrao Jan or by the author in these purchases because Khanum, who bought Ameeran, was kind and saved the girl from a worse fate by giving her shelter. Ruswa in no way questions the social system that allows the buying and selling of girls, although he is full of sympathy for the plight of individual victims. Umrao Jan also takes it for granted that there will always be respectable women and courtesans, and that there will always be an uneasy rivalry between them. Ruswa could not have intended to present his own character ironically, but to the modern reader there is an ironic gap between his pious lecture on the superiority of the mothers and wives in purdah and his familiarity with all the courtesans of Lucknow. Umrao Jan, after describing an incident when she was insulted by the ladies of Akbar Ali Khan's house, comments on the way women of her profession are treated: 'Mirza Sahib ... it is inhuman to look down upon another human being with such contempt.'[23] Mirza Sahib apparently does not think so and holds forth self-righteously. Umrao's gentle probing about the responsibility of men towards women is airily dismissed when he says that men love variety and need change. Ruswa does not emerge as a likeable character in the novel. He constantly prods Umrao to reveal more about the intimate aspects of her relationship with men, but she wards off his queries with urbane suavity, refusing to satisfy his prurient curiosity. They have a bantering relationship which is sometimes mildly flirtatious (p. 177), but Umrao cherishes it as belonging to that ideal category in which there is a demand for nothing other than good conversation.

Yet we are not meant to condemn Ruswa as a hypocrite. While discussing the nature of sin with Umrao Jan he quotes Hafiz: 'Take to drink or worship idols, burn the kaaba or the Koran. These the Lord might perhaps forgive, but not that you hurt a man.' This couplet has a special significance for Umrao Jan, who never hurt anyone—who only gave men pleasure and thus could hope to be forgiven by God.

The tension in this novel is between Ruswa's sympathy for the character he has created and the social attitudes he must uphold. The double standard is accepted when he absolves his own guilt about visiting brothels by declaring, 'Attempts to mislead innocent women cause me great sorrow, and if I had my way, I would put all those who try to seduce the virtuous against the wall...'[24] This is comparable to Bankim's attitude to Rohini, the young widow in *Krishnakanter Will*, where sympathy for her as a person has to contend with the moralist's insistence that the social order must not be disrupted.

The courtesan is not quite outside the social order—she has a role to play as long as she is young. It is the loneliness of the aging *tawaif* that gives the novel its special human poignancy. The mature Umrao struggles with a desire for emotional stability and the knowledge that she must learn to live alone. Unwilling to let any trace of sentimentality cloud her urbane and witty relationship with Ruswa, she deliberately takes a cynical and detached view of the man-woman relationship, offering a biological explanation for emotional states. For women's ability to simulate love she has this explanation: 'weaker animals are endowed with the ability to deceive in order to preserve themselves; and for women's preference for young men, this:

'being weak themselves,' they wish to provide themselves with protectors. Young men, being able-bodied, can be relied on more than old men in times of trouble.'[25] One is not sure if there is a touch of cynicism in her talk of the solace of religion and literature when youth is over: 'When they [her admirers] began to drop out of my life one by one . . . I developed a taste for books.'

The narrative proceeds from the time Ameeran is kidnapped to the time when she leads a pious and literary life in the seclusion of her home. Many years pass in between. The 1857 revolt rocks the country, Wajid Ali Shah is forced to abdicate, prosaic British rule replaces the charming eccentricities of the Nawabi ethos, the seat of power and influence moves from Lucknow to Hyderabad, the frightened Ameeran grows into the famous poet-courtesan Umrao Jan. How is the passage of time indicated? Sometimes through simple statements: 'I grew to adolescence... At fourteen I learnt to gaze at myself in the mirror', or, 'I had been in Kanpur for about six months'. Sometimes Ruswa helps Umrao Jan to fix dates by referring to public events. A sense of passing years is suggested when Umrao reaches Faizabad, where she spent her childhood, and measures the distance she has covered in terms of her own change: 'But I had lived through the tragic days of the Mutiny, seen kingdoms collapse before my eyes, witnessed the fall of princes like Birjis Qadr, and my heart has become as hard as stone.'[26] But nowhere is the passage of time more effectively conveyed than when, after many years, Umrao goes to her old room in Khanum's establishment where she began her career. The room has been locked for years. It evokes memories despite dust and cobwebs, and the centipede that crawls along her dupatta. The five gold sovereigns hidden under the bed—her first private earning—shine amidst the squalor, gathering up as it were all the intervening years and revealing the bright shine of a first love that had since been dulled through repetition.

Towards the end, the narrative loses its episodic structure and the loose threads begin to be tied up. All the forgotten and lost characters turn up like the five gold coins, and even the men who kidnapped young Ameeran are arrested and sent to jail. Thus the novel has two structures: a neat, cohesive, well-organized one where episodes are linked to each other, where the criminals are punished and Umrao, after a successful career, goes into a well deserved and comfortable retirement where the British come to make life orderly and peaceful; and the other, a drifting aimless story of a solitary woman without moorings, of one who faces loneliness at the end of a life artificially filled with laughter, music and protestations of love, of one whose childhood friend Gauhar deserts her after grabbing a large share of her fortune, and of one who turns to books and religion for solace because her life lacks the warmth of human relationship. The conflict between Ruswa's artistic insight into personal psychology and the awareness of his public function as a writter can be seen in these two simultaneous levels of apprehension.

Realism is not a value in literature, but one of the many modes that narrative fiction can adopt. When the novel was emerging as a distinct genre in India, social realism had for some time been the dominant mode in the European novel, and the early Indian novelists joined in 'that effort, that willed tendency of art to approximate reality.'[27] This effort consisted, among other things, in the creation of characters in situations permitting individual choice as well as their mimetic representation in a manner which did not distort contemporary Indian reality.

The challenge to achieve this successfully was intensified when the characters represented happened to be women. Social conformity has always been more obligatory for a woman than for a man, and generally a woman's identity tends to be defined, by herself as well as by others, in terms of her relationship with men—as a daughter, as a wife, as a mother. This was almost equally true for nineteenth-century European women as well. One of George Eliot's characters in *The Mill on the Floss* (1860) reflects a generally accepted belief when he says: 'We do not ask what a woman does. We ask who she belongs to.'[28]

It is not entirely an accident that several major European novels of the nineteenth century (*Vanity Fair, Middlemarch, Madame Bovary, Anna Karenina, Emma, Portrait of a Lady*) have women as their protagonists. Social realism at its best conveys in concrete and specific terms the complex relationships between individuals and their society. This relationship can be studied in sharper focus when the individual's life restricted within a narrow sphere where she is permitted very few options, and when the odds are against her, in other words when she is a woman.

But there is a basic difference between the woman protagonist of a nineteenth-century European novel and her counterpart in India. Individualism had been emerging as a human ideal in the West for a couple of centuries, and even though in actual fact a woman's life lacked the relative autonomy of a man's, the possibility did exist as an idea. In India even though exposure to the West was beginning to make an alternative ideal available in the nineteenth century, such individualism was alien to traditional thinking. In the four novels discussed above the choice of women protagonists might have been a deliberate device to magnify for closer analysis the conflict between the restrictive social norms and half-articulated yearnings to achieve selfhood. The design was not always clear cut or schematic. Each writer had to formulate his own strategy in tackling the problem. The resultant works reveal tensions of different kinds—between form and content (*Pan Lakshyant Kon Gheto*), between intention and execution (*Indulekha*), between imaginative sympathy and social responsibility (*Umrao Jan Adā*), between the author's conscious design as reflected in the plot and unconscious longing as revealed in language and imagery (*Indira*).

It is a critical platitude to say that the Indian novel has a derivative form, imitated from the West. This is only superficially true. A form cannot be superimposed upon a culture which lacks the appropriate conditions to sustain its growth. If the Indian writers did not come in touch with English literature at this historical moment, and if they did not accept realism as a desirable mode of narration, which direction their narrative aspirations would have taken remains a matter of speculation. In the nineteenth century fictional texts that are available to us we find that the reality of the Indian social situation was often moulded and manipulated to meet the exigencies of this new western literary mode. The realistic novel was able to come into existence because the tension between individual and society had acquired a certain intensity. Had this tension not existed, narrative fiction may have continued to retain qualities associated with the epic or the romance; and if the tension had become so acute as to threaten social disruption, creative expression may have turned inward, towards solipsism and away from realism—perhaps poetry would have been a more suitable genre in such a situation. Therefore if the social transformation of the nineteenth century had not set in motion certain dialectical forces among the English-educated class, the novel in its realistic form might not have taken root in India. This chapter has sought to isolate some of the attempts to reconcile social reality and fictional realism which mark the creative imagination in nineteenth-century India.

V
THE NOVELIST FOR ALL SEASONS

Singling out for discussion three different strands of early fiction in India—utilitarian exercise, historical romance, and realistic narrative—as has been done in the foregoing chapters, is at best a strategy for exploration. It cannot hope to sort out comprehensively all the skeins of a complex tangle. The tangle had regional variations, both history and geography complicating the knots. Certain coastal regions and other trading centres had access to foreign influence sooner than inland areas, while local history and the literary and folk tradition of each region left their imprint on the early prose narratives produced in the region. Literary output was also affected by political and economic changes. Punjab, for example, was one of the later areas to come under British administration; we find that Punjab held on to its folk tradition of heroic tales in verse longer than urban centres where the printed word and discursive prose had begun to take precedence in the late nineteenth century. So many factors distinguished the culture of one region from that of another that each regional literature began subsequently to think of its own development as a unique process, fit to be studied in isolation. Yet, as we read fiction down the decades and across the languages—even in incompetent translations—we sense that comparable elements exist in most parts of the subcontinent, that there are certain shared parameters in the literary culture of India.

Instead of searching further for more common strands in the development of the novel in India, I will adopt the alternative strategy of looking at the work of a fiction writer whose nationwide acceptance has made it possible to think of a pan-Indian readership of the Indian novel. Such a writer was

Saratchandra Chattopadhyay (Chatterjee) whose novels appeared to find no linguistic or other cultural barriers as they
entered homes in all parts of the country.

Saratchandra died in 1938 but is still very much a part of
the 'present' of Indian literature and likely to remain so for
years to come. No Akademi prize or Jnanpith award gave him
status, no recognition from abroad earned him his domestic
reputation. He must be the only Indian writer who needed no
official patronage to get translated from Bengali into other
Indian languages, is used as a point of reference in any literary
discussion anywhere in India, and continues to be read widely
long after he has ceased to be fashionable. All his novels are
available in almost every Indian language, though very little of
his work has been translated into English.[1] This alone should
tell us something about the extent of his appeal and the nature
of his readership—which indeed is more a reading republic than
a mere public. Irrespective of the fact that serious present-day
literary critics prefer to leave him alone, his highly durable
grassroots popularity is a phenomenon of continuing cultural
significance in modern India, proving adequately, if proof were
needed, that despite surface differences there is a common
Indian substratum of literary taste at the mass level.

In his lifetime (1876–1938) Saratchandra was considered
something of a rebel in Bengal, a challenge to orthodoxy and a
threat to the establishment. His novels were said to be sharp
strictures against the hypocrisy of caste-bound Hindu society
and its treatment of women. Novels like *Srikanta* and *Sesh
Prashna* even questioned hallowed concepts like *satitva* and
the sanctity of the marriage tie when love is absent.

Radical social attitudes and mass popularity generally do not
go together in the Indian context. It is well-known for example
that the commercial Hindi film—another popular all-India
phenomenon—uses the veneer of social consciousness (the
'bad rich' and the 'good poor') to perpetuate feudal, hierarchical and sexist values. This apparent paradox in Saratchandra,
the coexistence of radical thought and mass-popularity, may be
analysed by isolating two of the recurrent concerns in his

novels: the place of woman in the family and of caste hierarchy in society.

I

More than half of Saratchandra's twenty novels and about as many short stories deal either centrally or partially with the situation of a widow. This is not merely a device for evoking the reader's sympathy—it had a basis in the social reality. With the polygamy of the Kulin brahman in Bengal, early widowhood was a genuine problem in the nineteenth century, continuing to a lesser degree in the early twentieth. Many of these young widows never had a chance to know or love their aging husbands who visited them in their father's houses once a year to collect a customary payment. Prior to Saratchandra, Bankim (in the figure of Rohini in his *Krishnakanter Will*, 1876) and Rabindranath (in the character of Binodini in his *Chokher Bali*, 1903) had created two unforgettable widows whose vitality clashed with the rigid code of penance society enforced on them.[2] As we have seen earlier, the problem of the widow was central to Apte's *Pan Lakshyant* and to its remote ancestor *Yamuna Paryatan*. In a strictly structured role-oriented society the widow's position was an uneasy one. The unchannelised sexuality of a young woman without a legal owner posed a threat to the stability of orthodox society. The severe rules of conduct and attire that were imposed on them on spiritual or religious pretexts could be seen as society's attempts at desexualizing them so that the potential danger is minimized. The situation had not changed even in the twentieth century when Saratchandra began writing.

Abhaya, one of Saratchandra's characters (*Srikanta I*), rebels against the word *brahmacharya* as applied to the widow's code of conduct. When Srikanta chides her saying, 'Don't call it brahmacharya if you don't like the word. What is in a label?' Abhaya indignantly replies, 'Everything is in the label, Srikanta Babu. What is more real in the world than words? Don't you know how man's thoughts can be perverted by giving them a wrong name?' Another character wants to know why a woman's value should be measured only in terms of her ability to undergo privation. Even when they are not so openly rebellious, other widows in his novels dare to fall in love, knowing that their love cannot be socially sanctioned.[3] Saratchandra treats these characters with sympathy but it is curious that not once in his twenty novels is such love seen to culminate in marriage or any kind of fulfilment. Savitri in *Charitraheen* is made to surrender the man she loves to a woman socially more acceptable, Rajlakshmi does not want to degrade Srikanta by laying claims on him, *Pallisamaj* ends when Rama goes away to Kashi, and Neelima in *Sesh Prashna* removes herself from the scene to be quietly unhappy somewhere else. The examples could be multiplied.

Saratchandra's persistent refusal to confer happiness on a widow may be testimony to the failure of Ishwarchandra Vidyasagar's crusade for the remarriage of widows more than half a century earlier. Nevertheless, in view of Saratchandra's sympathy for and understanding of the need to look at these women as individuals rather than as part of an unfortunate category, this is disconcerting. That he avoids a happy ending may be interpreted in any or all of three ways. First, he was being faithful to the social reality of the time. A widow's rebellious remarriage might have been a romantic gesture but was hardly likely to succeed. In one impulsive moment Rama and Romesh might have decided to get married, but in which society would they have lived? Pallisamaj would not accept them. Besides, the barriers were as much within the characters as without. Secondly, unfulfilled love as a higher value than fulfilled love can be seen as part of a mythic motif in India–Radha's separation from Krishna being the archetype. Union in marriage is a limited goal compared to the transcendence towards infinity achieved through perpetual *viraha*. This idea exists at various levels in Indian culture, supporting the notion that suffering or sacrifice is nobler than fulfilment, renunciation greater than gratification. The last paragraph of *Srikanta I* contains a well-known sentence: 'Now I could see that the highest kind of love not only brings people together, it can also pull them apart.' Thirdly, these unhappy endings can be seen as

Saratchandra's concessions to the taste of his audience. Sympathetic understanding of a problem was one thing, accepting a drastic solution another. The middle-class reader was prepared to wax emotional over Srikanta's love for a childhood playmate who had been married, widowed and sold to a Maithili landlord and kept as his mistress, but a consummation of this love in marriage would have shocked his sensibility.

It is possible that Saratchandra did not make this concession consciously—his inhibitions might have happened to coincide with those of his readers. Saratchandra typified the ambivalence of the educated Indian middle class which has compassion for the exploited, but is reluctant to disturb the social order in which it has a vested interest.

This ambivalence may be the cause of his popularity but it often plays havoc with Saratchandra's characterization. Kamal in *Sesh Prashna* is built up as a nonconformist woman who argues about the meaninglessness of rituals and refuses, at least in theory, to be tied down to a relationship when love has disappeared. Yet in her own relationship with men she is found enacting a traditional role where cooking an elaborate meal for a man and nursing him in illness appear to be the final aims of a woman's existence, as if her real salvation lies in an extinction of selfhood. Saratchandra persistently upholds this role for his women and repeatedly sets the stage for men to fall ill suddenly or feel hungry at odd hours. Such recurrring scenes reduce the complexities of human nature, which are reduced to a formula and the individual woman is stifled by the feminine stereotype.

In the work of Premchand, Saratchandra's contemporary, the remarriage of a widow is not seen to be impossible. In *Godan*, Gobar marries Jhunia after getting her pregnant, and the family, after initial hesitation, accepts her as a daughter. This is possible because Premchand is not writing about the middle class, where inhibitions tend to be more rigid. But Saratchandra, even when he is not writing about the middle class, sees every character through the lens of middle-class values, investing each woman with the halo of a self-sacrificing goddess. Savitri in *Charitraheen* is the cleaning woman in a boarding house; Kamal in *Sesh Prashna* is said to be the illegitimate daughter of a British tea-planter and his Bengali housemaid. Yet in Saratchandra's world their virtue lies in their embodying values of the middle class. Kamal's unusual birth remains a gratuitous piece of information in *Sesh Prashna*, as do the unusual backgrounds of several female characters (Rajlakshmi, Savitri, Annada-didi). Unlike in Rabindranath's *Gora*, heredity and environment in no way determine Kamal's personality, demonstrating Saratchandra's inability to view an individual in its totality—in terms of spatial, social, cultural and economic axes.

'If I had five thousand pounds a year, I could be an honest woman', says Becky Sharp in Thackeray's *Vanity Fair* in 1847. In Saratchandra's own time, Gobar in Premchand's *Godan* says: 'If had two square meals a day, I too could become religious.' Saratchandra refuses to admit that middle class moral and ethical values can operate only at a certain economic level. When Kamal's husband deserts her she carries on in her noble sacrificial role. Mundane details such as the sources of her house-rent and food are glossed over by a casual hint that she makes a living by stitching clothes for slum children! The difficulty of being a 'good woman' without financial security is not something with which Saratchandra is concerned.

Saratchandra once proclaimed *manushyatva* (humaneness) to be greater than *satitva* (chastity), but a close analysis of his novels shows that he always saved his women from physical 'impurity', as though without chastity all humane qualities are nullified in women This often reduces his women into feminine stereotypes, but paradoxically this is also the source of the novelist's popular appeal. Because he had left the basic values undisturbed, he was permitted by his readers to critique certain other aspects of social behaviour.

II

There is far more complexity and contradiction in Saratchandra's handling of social hierarchy than in his treatment of women. While Saroj Bandopadhyay has shown, in an excellent analysis, that unlike Rabindranath who understood the historical forces disrupting the economy of rural Bengal and the organic unity of the villages, Saratchandra saw problems in a very fragmentary way,[4] the fact remains that Saratchandra knew the individuals who made up a village even though he may not have grasped the economic system. But for two exceptions (*Mahesh* and *Abhagir Swarga*, which, incidentally, are his best short stories), all the major characters in Saratchandra's work belong to the land-owning professional class who in Bengal happened to be the upper castes. (They called themselves *bhadralok*, as different from the lower castes and the poor who were called *chhoto-lok*). The novelist's central concern never shifts beyond the bhadralok. One may hear of Muchipara where the cobblers live but one is seldom taken there unless an epidemic breaks out, as it does in *Grihadaha* and *Sesh Prashna*. The bhadralok characters go there to nurse the poor and we are expected to admire their selfless nobility rather than look at the lives of the cobblers.

When Saratchandra wrote, the drift from village to city had began and the neglect of the village by absentee zamindars was one of the writer's regrets. The young ascetic Vajrananda's words to Rajlakshmi express the author's own views: 'The misery of the poor people has increased fourfold because you live away from the village. Not that you did not exploit them when you lived here, but what your absence is doing to them is much worse. When you were here, if you made them suffer, you also shared their sorrows. If the king dwells in his own kingdom, his people's cup of misery can never be so full.' (*Srikanta II*, p. 176) The regeneration of the village, Saratchandra felt, could come only through the efforts of the educated, liberal and kind landlords who would return to live in the village, as do Romesh in *Pallisamaj* and the reformed Jibananda in *Dena Paona*.[5] Saratchandra's faith was in the kindness of individuals rather than in collective social action. What George Orwell said about Dickens applies to Saratchandra if we change the industrial context to the agrarian: 'Nowhere does he make any attack on private enterprise or private property ... the whole moral is that the capitalists should be kind—not that the workers should be rebellious.'[6]

Saratchandra's blissful indifference to economic issues lends a touch of the fabulous or unreal to his handling of money matters. Money seems to be something one either happens to possess or not to possess; it is not recognized that the method of making a living must profoundly influence behaviour. And as in popular Indian films, Saratchandra's characters inhabit a world where happiness is entirely dependent upon an emotional adjustment with the other sex. The larger historical and economic forces and professional pressures that impinge upon private lives are generally ignored. Suresh (in *Grihadaha*) has a huge unearned income, therefore he can travel at will, buy a house anywhere he likes, furnish it as he pleases and equip himself with a carriage and two horses. Rajlakshmi in *Srikanta* has several well-appointed houses in Calcutta, Patna and Benares; she buys up a whole village halfway through the novel and another village by the end. She can not only travel whenever she feels like but can send her servant with 'a bundle of money' every time Srikanta happens to be short of cash. The conception of such convenient wealth smacks of childish fantasy specially when the earning or possession of this wealth seems to leave no mark on character. Some of Saratchandra's most sympathetically conceived characters are rich but indifferent to money, and recklessly give away or squander what they did not earn (for example, Devdas; Gahar in *Srikanta*; Priyo Mukhujye in *Bamuner Meye*).

Caste identity and religious faith serve as indices of human conduct in Saratchandra. Characters most rooted in these elements are the ones who are most convincingly alive. *Dena Paona* and *Bamuner Meye* are perhaps his most successful novels, capturing unerringly the tensions generated by the clash of different kinds of power and pride in the traditional structure. The novelist does not point his finger at the corruption from a distance; he knows it from the inside. In *Bamuner Meye* Golok Chatterjee can chastise Arun for having gone abroad and not performing the purification rite afterwards, but he himself gets his money by exporting goats and sheep. What is more, he helps a Muslim trader to export cows—the most damning fact about Golok both from the reader's as well as the author's point of view. Caste-proud Sandhya is unable to marry Arun not because society opposes this but because she herself recoils from a brahman of a lower order. When Sandhya hears of a misalliance that had defiled the purity of her own blood in some past generation, she faces an identity crisis. Loss of caste is for her true alienation. Shorashi in *Dena Paona* believes in her role as *bhairavi*—the dedicated priestess in the temple—and her conflicts arise out of her belief in this partial deification of herself and her ordinary instincts as a woman.

Caste is identity in Saratchandra's world, giving characters a fixed place in the social order. Therefore when the author uproots them from their locale he loses his grip over them. This removal is sometimes done to isolate them from the enveloping bondage of society and to study them as individuals. Achala urges Suresh (in *Grihadaha*) to take her to a place where there are no Bengalis, i.e. no familiar social fabric. The unorthodox behaviour of the characters in *Sesh Prashna* is made possible because the locale is Agra. When further escape is necessary, a couple disappears to a remoter place—Punjab. Burma sometimes serves a similar purpose, though Burmese society has a little more specificity in Saratchandra's novels than the region vaguely called 'paschim'.

In Burma Srikanta is made to realize how the rigidity of caste and pollution taboos becomes irrelevant when one is far from home. On the rational level he sees this as a beneficial change and comments on the necessity of coming out of the closed community of the village to realize the narrowness of its worldview. Yet his instinctive revulsion at the loosening of caste rules is revealed in the tone in which a squalid restaurant in Burma is described, where Indians of different castes eat together. Srikanta is shocked to find that in Burma brahman cooks clean utensils and polish shoes for extra money. He pauses to make scathing remarks about the mercenary nature of North Indian brahmans who stoop so low for money, and smugly declares that a Bengali or Oriya would not sell his brahmanhood so easily. (p. 132). Examples can be repeated to illustrate

Saratchandra's ambivalence towards the caste system and the purity and pollution laws governing it. When the pride of caste makes a man inhuman, Saratchandra castigates it. If women are sacrificed at the altar of caste, his condemnation of the rigidity is unequivocal. But when the observance of caste rituals maintains social order, he finds in this the analogy of a family. The fabric of a family or a closely-knit community is necessary for Saratchandra's characters to come alive. Both the joint family and the village community are, in his novels, paternalistic organisms containing at their best a harmonious coexistence of the parts in their proper hierarchical order, at their worst oppression and dehumanization. This implied family–community analogy comes to the surface in a passage in *Pallisamaj*. When England-returned Romesh frets about caste inequality in the village his aunt reminds him:

This is not your city my son. In the village no one worries about which caste is higher, which lower. Just as a younger brother does not mind being born a few years later and does not feel ashamed to touch his elder brother's feet, so the kayastha does not hesitate to touch the feet of the brahman. (Vol. I, p. 423)

The evil, she suggests, lies not in the system but in its corruption. The joint family offers security in return for a surrender of ego. It is not allowed to disintegrate into nuclear fragments in *Nishkriti* and *Ramer Sumati*.

In *Srikanta*, Sunanda, a spirited woman who rebels against exploitation, is finally condemned because her 'bookish' theories of right and wrong break up the joint family. The joint family in Saratchandra's world is a benign institution which protects even those who are weak and impractical, exactly as a close-knit village community supports a few wasters. In *Srikanta III* the narrator has a long conversation with a couple of enlightened villagers about the new mercantilism and the consequent corrosion of human values in villages. One of them says:

You see there have always been in our villages a few men who were indifferent to money, who lacked material enterprise and spent their lives in unprofitable work. They played chess or games of dice at the grocer's or the sweet vendor's shop; they helped to cremate the dead:

they sang and played music at the mahfils in rich peoples' houses; they organized the collective festivities and public rituals of the community. They did not always have enough in their own houses, but managed to live on the surplus of others. You English-educated people have contempt for them. But don't worry, these parasitic and useless people have now disappeared ... but with them the joy of our villages has also vanished. (pp. 224–5)

Saratchandra looks back with nostalgia towards a vanishing world where communities were organic and self-sufficient, where individuals were linked with others in a clearly defined bond.

In Saratchandra's novels characters are forever establishing family relationships even with strangers and outsiders, thus appealing to the nostalgia of readers who are under the stress of a gradual transformation both of the family and the community. Individualism that can sustain urban life is alien to agrarian Indian values, and Saratchandra's irresistible appeal lies in his upholding an integrated life that is daily becoming unattainable. Perhaps this is also the reason why he appeals less to the urban intellectual or second generation metropolitan reader than to those who are still emotionally attached to land and the expansive bonds of the extended family.

An extract from Saratchandra's *Pandit Mashai* probably sums up the entire argument: that one has to belong to the structure in order to criticize it. Criticism from the outside seldom achieves the same validity. In *Pandit Mashai* two friends, Keshav and Brindavan, have both been trying to run schools for poor children in their respective villages. Brindavan's school thrives while Keshav complains that he gets no students. Brindavan tells him:

'If you reject the beliefs of the village people as superstition, how will you make them trust you? Tell me Keshav, as a brahman do you perform your daily rituals?'
'No.'
'Do you keep your shoes on while drinking water.'
'Yes.'
'Do you eat food cooked by Muslims?'
'I might, I have no prejudice.'
'Then let me tell you, your decision to run a village school to teach

the poor children is an exercise in futility. Or more than futility, if you don't mind my saying so.'
'Presumption?'
'Exactly. Keshav, it is not enough to have a kind heart and willingness to help people. You must be able to share the life of the people whom you want to help. If you go too far ahead of them in your knowledge and belief they will not understand you, nor will you be able to reach them.'(Vol. II, p. 318)

Saratchandra reached such an enormous number of readers because he did not go too far ahead of them, because, metaphorically, he knew he had to remove his shoes while drinking water. When he pointed out corruptions and inhumanities in society he could make an impact because he belonged to the society he was criticizing.

Saratchandra was a popular writer par excellence and used various situations and motifs that have become formulae since his time. The Bombay film has learnt from him these devices and has managed to squeeze the maximum possible mileage out of them—the idealized woman wronged by society but noble in suffering; the unrequited lover who drowns himself in drink; illness as *deus ex machina*; the isolation of protagonists from the forces of history and economics; the establishing of filial and sibling relationships with complete strangers; the continued intensity of pitch and emotionalism. Most of these familiar elements of the entertainment industry can be traced back to Saratchandra.

But his novels remain as different from these products as folk music from pop music—the one is spontaneous while the other is manufactured synthetically with an eye on profit. Saratchandra's intentions were far from commercial. He was inspired by a genuine interest in human beings and a passion for exposing social injustice. This coexisted with a contradictory impulse to maintain order and discipline in society. This paradox was essentially what made him popular because it reflected the contradictions inherent in his extensive middle-class readership. His popularity continues at some levels because of certain unchanging aspects of Indian middle-class values.

III

The four-part *Srikanta* (*I*, 1917; *II*, 1918; *III*, 1927; *IV*, 1933) is worth looking at in some detail because it spans almost the entire writing career of Saratchandra and contains in essence all the structural motifs, cultural ideals, moral values and social attitudes that mark his work. This tetralogy is picaresque in mode and autobiographical in technique—and to an extent in content as well. Like Srikanta, Saratchandra spent his childhood in a Bengal village, his adolescence in Bihar, worked as a clerk in Rangoon and generally led a rootless, wandering life. But to trace correspondences between his life and his fiction is not necessary for our analysis.

The 'tramp' hero—the epithet aptly commemorated by Vishnu Prabhakar in the title of his Hindi biography of Saratchandra, *Awara Masiha*—is provided by the narrator himself in the first sentence of the first volume. This narrator encounters during his peregrinations diverse characters and strange situations which are recorded in an episodic manner in the four volumes. The only character who appears in all the volumes is a woman called Rajlakshmi to whom Srikanta is bound by a fluctuating emotional tie. Continuity lies also in the manner of perception, the curiously detached emotionalism with which Srikanta responds to experience.

Two premises are set forth on the first page of the work: one, the narrator stands outside the normal expectations of organized society; two, his narration is a bare and unembellished report of actual events. The second however is merely a device to ensure credibility because the tone of the novel is hardly that of reportage. The language is generally charged with sentiment; in the first volume there are two famous passages of evocative prose—one describing the cremation ground at night and the other a cyclone in the Bay of Bengal—which stand out brilliantly purple in their literariness. Yet on the very first page the narrator disavows 'literariness' by declaring that he is not a writer:

Travelling is one thing, writing about it quite another. Anyone who has legs can travel, but not all who have hands can write. Writing is not easy. And unfortunately God has not blessed me with the slightest trace of poetic imagination. My wretched eyes see everything exactly as it is. A tree looks like a tree, the hills remain mere hills. A stretch of water never seems to be anything more than that. I have developed cricks in my neck by looking long at the clouds, but have not been able to detect even a strand of raven hair in them, not to speak of long tresses. I have never seen a fair face in the moon even though I have gazed at it till my eyes ached. One who has been so deprived by God cannot create poetic images. He can only report the truth plainly. That is what I propose to do. (p. 1)

Soon after this preamble the reader is introduced to the one remarkable male character in the tetralogy (all the other memorable characters are women)–Indranath–Srikanta's mentor and friend during the impressionable years of adolescence. It was Indranath who initiated Srikanta into a life of non-conformity, steering him away from the norms of the acquisitive, orthodox society. Indranath is a hashish addict. He beats up people, he steals in order to help others, he can row a boat recklessly in a stormy river; but he can also play the flute hauntingly and be very tender. All these add up to a composite figure that has almost the dimensions of an archetype. At some subterranean level in Indian culture there is great admiration for a man who is indifferent to material values, for one who is at the same time without guile and subterfuge. These qualities somehow exempt him from the need to 'be good' in the usual sense of the term and, although society may not exactly condone his excesses, his *badnaam*[7] life is regarded with relative indulgence. The unfailingly romantic aura of such a man is not Saratchandra's creation, but he did a great deal to popularize the concept of the unworldly vagabond who loves and leaves, drinks and debauches, and is forgiven. Srikanta himself is one version of such a figure. But before he develops into a character, we have a brief but memorable glimpse of the archetype in Indranath.

Indranath is a more dynamic character than Srikanta, even though he is never seen after the first forty pages of this long tetralogy. Since he combines a fearless adventurous spirit with a streak of renunciation, he is the one complete ideal against which the other fragmentary male characters are to be judged.

Buddhadeva Bose once pointed out that Saratchandra is at his best when his characters are 'chronological adolescents instead of chronic ones. For in growing up they threaten to outgrow their author and he hardly knows how to deal with them.'[8] Perhaps the strength of Indranath as a character lies in the fact that he does not have to grow up.

Compared to Indranath, Srikanta is a passive figure who drifts here and there without much premeditation or purpose, often overshadowed by the women who crowd his life. At least five of these deserve special notice, one from each volume—Annada, Abhaya, Sunanda, Kamal-lata, and Rajlakshmi who dominates all four volumes. Only two other men are of more than casual importance in the entire tetralogy: Vajrananda, the young ascetic who appears in Part III, and Gahar, the village poet who dies soon after he enters Part IV. On the whole it is the women who give the whole work its density and variegated texture.

Swami Vajrananda is one of the few men who undertakes disinterested social action in the entire corpus of Saratchandra's novels—two others being Brindavan in *Pandit Mashai* and Romesh in *Pallisamaj*. Although Vajrananda wears the sanyasi's clothes his devotion is not to God but to his country. He is reminiscent of Rabindranath's Gora in his earnest identification with the people and his attempt to understand rural India, but he is never faced with Gora's dilemma. Rajlakshmi tries to deflect his path back to the security of family and domesticity with the usual weapons of Saratchandra's women—food and affectionate fuss—but Vajrananda's commitment is firm. A medical doctor by training, he roams from village to village tending the sick, helping to set up schools, attempting to arouse political consciousness among the villagers. The only time any reference is made in *Srikanta* to larger historical issues is when the narrator involuntarily gets involved in helping the victims of a cholera epidemic in a railway colony and on the way back meets some of the disciples of Vajrananda who discuss the British impact on rural Bengal. An extended quotation may be necessary here to understand Vajrananda's, and through him Saratchandra's, views on colonial economy. The villagers ask Srikanta:

'What was the need, you tell me, sir, to pierce our land with another rail road? Do the people want it? They don't. Still, it has to come. We have no pond, no tank, no well; there is not a drop of drinking water anywhere; the cattle die of thirst every summer ... Cholera, malaria and all other kinds of disease are wiping out the population. But who cares? Our rulers are only interested in bringing the railroad to the interior so that they can squeeze us dry and take out all the grain for export. Isn't it true, sir? ... I grew up ... in a village where there was no railroad within forty miles. It was a time of plenty and food was really cheap. Whatever grew in one's garden was shared among the neighbours. Today no one wants to give even a bunch of bananas or a bundle of spinach to another. They are more keen to hand everything over to wholesellers who will take it away by the eight-thirty train.

Giving away something is now considered to be a waste. It is sad, sir, sad, how men and women all have become mean in their pursuit of money ... And the root of all this evil is the railroad. If it did not branch out in every corner of our land like veins in the human body, if exporting food and making money had not become so easy, our country would not have been reduced to this. (pp. 223-5)

This is comparable to Dickens' attitude towards the railroad and the so called progress it brought to the English countryside in the first half of the nineteenth century: 'the yet unfinished and unopened Railroad was now in progress; and from the very core of all this dire disorder trailed smoothly away, upon its mighty course of civilization and improvement ...'[9] Yet the comparison cannot go very far because Dickens was responding to the real contradiction of the machine age—the power for life coexisting with the power for death. The new social and economic forces of his time and of his land found their organic culmination in the building of the railroad, whereas for Saratchandra's villager it was a meaningless superimposition without any vital link with the life of the people. Thus it was seen as an agent of dehumanization rather than growth. This may not be entirely the author's point of view because this statement quoted above is followed by a half-hearted debate between Srikanta and the villager about the relative merits of individual enterprise and the collective welfare of the community. Although Swami Vajrananda is not present during this conversation his views are reflected in those of his disciples.

In polar opposition to the public-spirited Swamiji stands Gahar in Part IV, the self-absorbed village-poet who lets his property go to seed while he composes a new Ramayana. We are told Gahar belonged to a family of Muslim fakirs and his grandfather was a famous singer. Gahar's father gave up the traditional profession to make money and buy property with profits in the jute trade. Gahar, instead of acquiring his father's business acumen, inherited the talent of the grandfather. He was Srikanta's childhood friend, the first person to teach him the use of a shotgun. Since he also wrote poems he embodied to a lesser extent the irresistible combination of masculinity and imagination that one came across in Indranath earlier. But by the time Gahar appears as an adult in Part IV he has given up shooting birds and lives in a dream world of imagination, while his property is being encroached upon by others. Srikanta's attitude towards him is one of amused pity. The impractical man, one who gives away his unearned assets carelessly, has always had the sympathy of Saratchandra. Gahar is not much of a poet; the futility of his poetic enterprise is evident, but in the pattern of the novel he is important in being the third variation on the theme of the ascetic ideal, the other two being Indranath and Swami Vajrananda. The Sufis and the Vaishnavas of Bengal have always had a streak in common, and Gahar is a typical Muslim example of Saratchandra's recurring unworldly renouncers.

It is not an accident that almost all the major characters in *Srikanta* including the narrator have a tangential relationship with society, standing outside the pale of its structured and ordered network. Srikanta is a tramp, Indranath a rebel, Vajrananda a sadhu and Gahar a dreamer. This is even more obvious in the women. Annada-didi is a Hindu woman living with a Muslim snake-charmer; Abhaya sets up house with a man not her husband; Rajlakshmi is a professional singer who amassed a fortune by entertaining her rich clients. These are perpetual outsiders who have no inherited code of values, who have to wrestle with reason and instinct constantly to devise their individual moral systems.

Annada-didi (Part I) leaves the most indelible mark on young Srikanta's mind and for a long time he thinks her the supreme ideal of womanhood. Not many people knew that the Muslim snake-charmer was really her own Hindu husband who had disappeared after committing a murder. After many years Annada sees this truant husband on the street and leaves home to accompany him. In Srikanta's words, 'one whose real place was with Sita, Savitri and Sati, was known to the world as a fallen woman'. Srikanta's lyrical rapture is based on the fact that Annada placed her true satitva[10] higher than her social reputation. Not many readers can share the narrator's gushing admiration for this woman in view of the fact that the object of her devotion is not only a drunkard and a tyrant but has been a murderer and rapist as well.

After his encounter with Abhaya (Part II), Srikanta begins to revise his ideas about the supreme feminine virtue of satitva. Abhaya had come to Burma in search of an absconding husband whom she located but found difficult to live with. Not only had this husband acquired a Burmese wife and children, he was cruel and violent. A physically as well as emotionally battered Abhaya then decided to set up house with another man who needed her more. Srikanta is repelled by this act, especially because the memory of Annada-didi is still fresh in his mind. More articulate than most of Saratchandra's long suffering women, Abhaya challenges Srikanta with a few blunt questions such as why chastity should be the sole responsibility of the woman in a marriage, while the man is free to violate marital norms? Abhaya also demands to know why the ability to suffer should be equated with virtue in a woman. And finally, 'If a cruel, corrupt and debauched husband turns his wife out of the house, should that be the end of her life, the crippling of all her female aspirations?' (p.125). Srikanta's hesitant answer is indicative of the basic conservatism of his attitude towards women tempered by a newly forming understanding of Abhaya as an individual: 'You may be innocent in the eyes of God and he will bless you but man cannot enter the minds of others. Society cannot judge people in terms of what is in their heart. If you have different rules for each individual our social order will crumble.'(ibid). Srikanta wants to condemn this illicit liaison but gradually sees that passing moral judgement on others is not easy. Finally, when during an epidemic of plague Abhaya nurses Srikanta back to health with affectionate care, his admiration for her as a human being eclipses all prior moral stricture.

Sunanda in Part III stands outside the galaxy of Saratchandra's exploited women in her triumphant assertion of individual spirit. She is different from Annada, Abhaya, Rajlakshmi and Kamal in a number of ways. She is not outside society. She is happily married and unlike the other women is a mother as well. Her problem does not revolve around any man-woman relationship; she rises above the personal to take a stand on a social issue. Before Sunanda is introduced in person, the reader is made to hear about her rebellious spirit from other sources. She and her husband left the prosperous joint family to which they belonged when she discovered that their prosperity was based on the exploitation of one particular family of weavers. Like her husband, she too had studied Sanskrit. Together they set up house in a small hut and lived in penury by giving lessons. Srikanta and Rajlakshmi, when they actually see Sunanda, are impressed by her dignity and poise and her firm adherence to principles. Rajlakshmi starts visiting her regularly to learn Sanskrit and to be initiated in the path of spiritual growth.

This is the beginning of the estrangement between Srikanta and Rajlakshmi. Her religious craving and love for him are contradictory claims and, under Sunanda's influence, she leans heavily towards religion. Much later, in Part IV when Rajlakshmi and Srikanta have another reconciliation, she blames their rift on Sunanda:

'There is something seriously wrong in her learning. She cannot make anyone happy ... Her elder sister-in-law is not educated; but she has compassion for others ... Sunanda exposed the dishonesty of her husband's elder brother before the whole world. Is this what the study of scriptures teaches you to do? Until her bookish learning gets tempered by the understanding of human nature, the sorrows and happiness of people, their temptations and vices, her abstract principles of justice and duty instead of bringing good to people will only succeed in hurting them'. (p. 336)

The retracting of the author's position is evident. The independence of spirit that seemed admirable at first turns out upon closer analysis to be too rigid and devoid of human warmth. Sunanda's unbending adherence to abstract principles causes sorrow to those around her and breaks up the joint family—an unforgivable sin in Saratchandra's world. It is as if Saratchandra does not know what to do with this fiery character after having created her. Submission of self and not assertion of self has generally been the quality he admires in women, and Sunanda turns out to be too independent. So she has to be kept in the background and later condemned indirectly through the influence she exerts on others. Judged as a human being Sunanda might score high, but as a woman she falls short in Saratchandra's scale of values. Srikanta, like his creator, makes no secret of his belief that men and women should be judged by different standards: 'I would like to tell you that women are not men. Their behaviour cannot be weighed in the same scale, and even if that could be done, it would not be beneficial.' (p. 126) When in Part II Abhaya questions Srikanta on his double standards, he avoids a direct reply by suggesting that women achieve greatness through suffering and not rebellion: 'Grief is not privation, nor is it an absence. The grief that is devoid of fear is something that can be savoured as much as happiness.' (p. 126).

Rajat Ray has seen in this attitude something indigenously and authentically Bengali and has tried to track its origin to medieval Vaishnava poetry which moulded Bengali concepts of man-woman relationships through the Radha-Krishna myth. Radha was separated from Krishna when he left Brindavan for Mathura, never to return. She sublimated her grief at the separation (*viraha*) through a merger with God. Ray observes, 'Sorrow is its own reward, because it is the deepest realization of one's self. This, then is the core of *karuna rasa* and its difference from tragedy is fundamental. It has no element of terror, nor truly of pity: pure grief, because it is a means of union with God, is sublime. Paradoxically, such grief contains joy: the highest emotion cancels these distinctions of ordinary experience.'[11]

The influence of the Vaishnava cult is much more obviously present in the last part of *Srikanta* where the hero falls in love with a vivacious Vaishnavi called Kamal-lata. The women of this sect who dedicated their lives to Krishna occupied a special place in the literature of Bengal mainly because of the relative freedom in their life style. Tarashankar Bandopadhyay's unforgettable Vaishnavi, Rai-Kamal (in *Raikamal*, 1934), was almost a contemporary of Saratchandra's Kamal-lata (in *Srikanta IV*, 1933), and the literary ancestry of Sudhin Ghose's Vaishnavi heroine Mynah (in *The Flame of the Forest*, 1957) can perhaps be traced back to both of these. The protagonist-narrator of Sudhin Ghose's English novel (which is the last volume of a tetralogy)[12] found himself irresistibly drawn to this wandering ascetic, the girl 'with a swing in her hips and magic in her voice'. The narrator wondered:

'What would happen', I asked myself, if at the end of the kirtan singing on a moonlight night Mynah said 'Come with me'? What would I do if she proposed 'you will play on the flute and I shall sing and we shall move on…'[13]

This is exactly what they do at the end of the novel. Together they wander from place to place, Mynah singing the praise of Krishna and he accompanying her on the flute.

The Vaishnavi cannot settle down in one place like common householders and *Srikanta IV* ends with Kamal-lata moving on towards Brindavan alone and empty-handed. Like the first part of *Srikanta* the fourth and final part also ends with a separation, reiterating the essential picaresque pattern of the novel where all unions are temporary and incidental.

The novel is not about rooted people living in a stable and well-organized society, but about rootless persons living on its fringes who are forever moving from one place to another, looking for elusive goals. Only Rajlakshmi keeps returning and attempting to domesticate the tramp hero. In spite of the author's obvious sympathies, Rajlakshmi is the least convincing of all the characters in the tetralogy. She was Srikanta's childhood playmate. He rediscovers her when she has become a courtesan; that one chance meeting changes her life and she turns into a chaste Hindu woman whose life is thereafter spent in mothering her step-son, tending Srikanta during his illness and generally feeding whoever happens to be hungry. References are constantly made to her childhood love for Srikanta—something she must have sentimentalized out of all proportion because he does not seem to have any clear recollection of the episode. The talent for singing with which Rajlakshmi earned her great fortune remains an extraneous part of her life, something to be ashamed of, while fussing over the food and sleeping arrangements of Srikanta becomes her main preoccupation. One wonders if concern about food and physical comfort does not become a sex-surrogate in Saratchandra's novels, for carnal passion is curiously absent. Rajlakshmi makes Srikanta's bed with her own hands, tucks him in with care and most of their conversations take place on or around a bed, yet physical intimacy is never even hinted at.

Rajlakshmi is the embodiment of all the virtues Saratchandra glorified in women: self-sacrifice, religiosity, compassion and sentimentality. If she does not become a living character the reason must lie in Saratchandra's own inconsistent attitude towards this character. Srikanta prides himself on his detach-ment, yet at all critical moments of his life seeks Rajlakshmi's help. When she begs him for a more permanent bond his answer is that he can bear every loss in life but cannot bear to lose self-respect, which marrying a courtesan would entail. Yet he does not mind when at the end of the second part she arrives in his village to nurse him during his illness. From the third part onwards Srikanta becomes a parasite on Rajlakshmi and there is a hint in the last part that after Kamal-lata leaves him he is going back to Rajlakshmi to settle down to the role of a permanent invalid so that in tending him Rajlakshmi's life gets a sense of purpose. More than one critic has observed that Rajlakshmi outlives her utility in the novel and weakens the total structure. Since, either due to the author's or the readers' prudery, passion has to be ruled out of their relationship, she ends up being the agent of Srikanta's emasculation, reducing his already passive figure to one worthy only of pity and commiseration.

Whether the weak and passive nature of the hero contributes to or subtracts from Saratcharandra's popularity could be an interesting speculation. A really dynamic protagonist like Rabindranath's Gora never appealed to the common reader's imagination as much as the ineffectual heroes of Saratchandra—like Devadas or Srikanta—whose prototypes continue ro reappear in popular fiction and films even today. Yet this 'softness' in Saratchandra's novels also alienated a certain kind of reader. Recently a Marathi critic has argued that although Saratchandra has been translated and read widely he has not influenced subsequent literature in Marathi because

the Marathi mind responds in a different way to emotional situations. It is not easy to enumerate the causes behind this, but maybe the history and geography of Maharashtra, its socio-economic conditions...have moulded the Maharashtrian mind in a different way... The Marathi reader is used to the overtones of intellectualism and reasoning. He is a little averse to the power of intuitive approach to truth...he cannot much relish the idea of understanding life through the medium of compassion.[14]

When Premchand used a similar argument in his letter to Jainendra Kumar to explain his aversion to the emotionalism of Bengali literature, it is very likely that he too had Saratchandra's novels in mind.[15] Saratchandra was never a writer's writer. Even in Bengali very little of his influence can be found in the serious literature of subsequent periods. Neither in the ruthlessness of Manik Bandopadhyay's existential vision nor in the pristine wonder and lyricism of Bibhutibhushan Bandopadhyay's evocation of reality can one detect any trace of the solidly middle-class sentiments of Saratchandra and his nostalgia for a certain social and familial structure. Tarashankar Bandopadhyay, a major Bengali novelist from the thirties onwards, was another self-tutored writer like Saratchandra, and equally unaffected by western literary norms. But his panoramic presentation of agrarian Bengal depended on a hard-headed understanding of the socio-economic reality of the caste and class structure of the time, and his novels are totally devoid of sentimentality.

Saratchandra's abiding influence is to be seen not in literature but in the realm of mass culture, particularly in that category of the entertainment industry called the 'family film'. While the action film and the sex-and-violence movies tend to be similar in different parts of the world, unique to the Indian subcontintent is a kind of emotion-charged domestic drama where the benevolent hierarchy of family and social ties is reaffirmed with emotional extravagance. Without being aware of his pan-Indian appeal, Saratchandra thus instinctively hit upon that elusive but pervasive element in Indian culture which is the most difficult to define, yet which distinguishes the instinctive responses of most common Indians from that of people of other countries and cultures. Hence any literary evaluation of Saratchandra cannot quite be delinked from his cultural significance, and when the literary critics have finally given him up perhaps the sociologist will take up for further scrutiny the phenomenon that is Saratchandra.

PART TWO

VI
PATHER PANCHALI

Forms of fiction are more elusive for purposes of categorization than poetry, though theorists of the novel—from Lukacs to Frye to Scholes—have attempted their own typologies. These theoretical discussions, in spite of their highly abstract criteria, tend to be culture-specific, pertaining generally to the literature of the western world. Some factors not generally taken into critical consideration may have to be introduced into the discussion when the novel to be analysed falls outside that cultural parameter.

More than fifty years ago Lukacs suggested a taxonomy of the novel depending to a large extent on whether or not the chief protagonist's soul is narrower or broader 'than the outside world assigned to it as the arena and substratum of its actions.'[1] The first category, for which Lukacs takes *Don Quixote* as an example, is marked by what he calls 'the hero's complete lack of any transcendental space', a description that fits many subsequent comic-satiric novels in the West, including Sterne's *Tristram Shandy,* and perhaps recently V.S. Naipaul's *A House for Mr Biswas. Pather Panchali* (Bengali; 1929) might be seen as an apt example of the other kind where the problem lies in 'the soul's being wider and larger than the destinies which life has to offer it'. It has the lyricism and sadness of the novel Lukacs quotes as an example of this type—Jens Peter Jacobson's *Niels Lyhne*—though not its disillusionment and desolation. 'Everything that happens may be meaningless, fragmentary and sad, but it is always irradiated by hope or memory'.[2] As in *Niels Lyhne*, hope and memory are important threads of *Pather Panchali* but 'disintegration and formlessness', seen by Lukacs as dangers inherent in this form, cannot be seen as such in *Pather Panchali* because its organizing principle depends on a totally different concept of time and reality.

In recent years Robert Scholes has suggested a generic approach to fiction through a convenient graphic scheme.[3] In trying to place *Pather Panchali* on this diagram we encounter various difficulties, because by a process of elimination the novel ends up being placed near the axis labelled sentiment, tragedy and romance. Each of these terms has such specific connotations in the history of European literature that it is difficult to use them out of their historical and cultural context. Seen from another point of view, *Pather Panchali* has all the elements of a bildungsroman so common in European fiction in the nineteenth century, or more specifically of a *kunstlerroman* because it records the birth, growth and education of its writer protagonist (but lacks the linear time-consciousness that sustains such a form). The concept of time here is cyclic and regenerative—giving an inclusive quality to the consciousness of its protagonist Opu. Seen by the standards which are applied to the 'well-made' realistic novel of the nineteenth-century western tradition, *Pather Panchali* lacks a cohesive structure, often meandering off in the manner of oral story telling. Opu's imaginative response to the external world is the only loose thread connecting the whole. In fact its form has been found so unwieldy by some that at least three European translators have tailored the novel to suit their idea of formal acceptability. It will be my attempt to explore the issues raised in evaluating a novel that does not quite satisfy the Eurocentric expectations of what a novel should be, yet remains a classic in its own cultural context.

Pather Panchali and its sequel *Aparajito* (1931),[4] which form a two-volume narrative tracing the growth of a village boy Opu, came to be known outside Bengal only in the sixties when Satyajit Ray made three films based on parts of this narrative. The stature of the original work by Bibhutibhushan Bandopadhyay (Banerji) however is in no way dependent on its celebrated cinematic transformation and its place was secure in Bengali fiction long before the film was made. The pattern of the narrative can be seen as a series of concentric circles, each expanding into the next larger, in the manner of ripples caused by a pebble thrown in water. At the beginning of the first volume the child Opu's horizon is limited to the courtyard of his house and the bamboo grove beyond it. Gradually his awareness grows to include the entire village Nishchindipur, then goes on to the fields beyond and to the next village, and by the end of the first volume he moves out of the boundaries of Bengal to go and live in Banaras. In the second volume, *Aparajito*, the expansion of his geographical and imaginative horizons continues until he leaves India from the port of Calcutta on a ship bound for South America. It is interesting that his destination should be not England or Germany but the farthest point that the Bengali imagination could reach in the early decades of this century.

Instead of leaving behind one phase of his life and proceeding to the next, Opu is able to incorporate each phase into the next one. Even as his life moves forward the details of his childhood do not simply fade away and drop out of his life. The meaning of the novel is not conveyed by the sequence of events but by clusters of images and states of mind that recur at different points of time. Sometimes a new experience echoes an older one (Durga, Gulki, the old brahman kathak in Banaras—all arouse the some protective tenderness in Opu) sometimes the old images are re-lived in memory in a new situation (feeling claustrophobic in the city Opu dreams of the local deity of his village—Vishalakshi—which echoes an earlier reverie of childhood), and sometimes the author himself steps in to connect the present moment to a future not yet known to the reader or to Opu. This aspect of contiguity is best illustrated in a passage at the end of the second section of the first volume. It occurs at a crucial moment in Opu's life when he is leaving the village of his childhood not to return for many years. As the train steams out of Nishchindipur Opu silently speaks to his dead sister, vowing that unlike the others he had not forgotten her:

It was true. He had not forgotten and he did not forget. Later in his life when he become so well acquainted with the earth and its girdle of oceans and tresses of blue; when his whole body thrilled to the speed of movement; when from moment to moment, as he stood on the deck of a ship at sea, the unearthly beauties of the blue sky flashed new on his sight; when the blue slope of a mountain wreathed in the vineyards faded from distance to distance and vanished beyond the dim bounds of the ocean's horizon; when the sweet siren melody from some far off shore, faintly discerned through a concealing haze, came to his ears like the voice of the lord of song; then and at all times like them, his memory took him back to a stormy monsoon night, to a dark room in an old house and the ceaseless noise of the rain, when the daughter of a poor village family spoke to him from her bed of sickness and said, 'Opu, when I get better, will you take me to see a train?' The distant signals of Majherpara station became fainter and fainter, then finally he could see them no more.[5]

In this passage the future and the past are fused together in the present. Opu leaves India to become in due course better acquainted with the ocean-girdled earth only at the end of *Aparajito*, but at the moment of parting from Nishchindipur the author projects the novel forward to another stage of his journey and backward to the memory of Durga, emphasizing the contiguity of all events, the irrelevance of the sequential development of time.

In *Pather Panchali* and *Aparajito* the seasons change, child-hood passes into adolescence and youth, but time never seems to be sharply fragmented. At the end of the two-part novel there is a kind of circular return to the beginning, when Opu's childhood is re-enacted through his son Kajal's response to Nishchindipur. 'The child's immediate and percipient penetration of the very sense of things'[6] is situated in a kind of a historical timeless continuum; Opu's childhood blends with his son's and while Opu leaves the country the cycle of growth continues in Nishchindipur. This circular movement of the novel has sometimes been regarded as formlessness, and in one case the English translators of the novel have even tried to justify why they felt the novel needed to be truncated. In the Bengali original of *Pather Panchali* the first section (six chapters) deals with incidents prior to Opu's birth; the second and the major section (twenty-three chapters) describes Opu's childhood and boyhood in Nishchindipur; the third section (six chapters) takes Opu out of the village to a larger world. This last sec-tion has been omitted in the English translation titled *The Song of the Road.* In the Introduction the translators argue that this part is extraneous to the requirements of the novel's formal integrity and imply that the exclusion improves the work of a writer who was essentially an untutored genius.[7]

But Bibhutibhushan's narrative is conditioned, perhaps unconsciously, by a very different concept of time and of aesthetic form than that which underlies western realistic fiction. The word *panchali* of the original title refers to a devotional narrative song which continues for a long stretch of time without any obvious climax, and where the story does not follow a rigidly sequential order. Quite often puranic stories and other tales in the oral tradition have this cyclic form wherein the events come back full circle to the initial state of equilibrium without registering any perceptible change on the axis of time. We may recall in this connection Mircea Eliade's distinction between mythic time and historical time—one being circular in movement, the other linear. The linear concept can be related to historicity, also with realism of the western variety, and a certain formal structure of the novel. In mythic or eternal time a cosmic rhythm embraces man and the universe in a cyclical repetition. The seasons of the year are an aspect of this principle of renovation and the renewal of generations is also part of the same process. It will be simplistic to assume that a novel written in the twentieth century can embody in its totality the time structure of myth or folk narrative, but *Pather Panchali* often reminds us of a rhythm of time that is far from linear.

Chapters in the novel are noticeably punctuated by reference to changes of season, and Opu's growing up is marked in terms of the summers, rains and winters he goes through. The festivals, so closely connected with the change of seasons, are also recurring measures of time. The autumn festival of Durga Puja marks the clearing of skies after months of heavy rain, and the Charak festival is associated with the onset of summer. For village children in Bengal these two festivals are the highpoints of the year and all the important events in the novel are seen in relation to these events. Durga dies during a particularly wild season of storm and rain. As the first tumult of grief dies down the skies clear after the monsoons. The sad-sweet music of Durga Puja can now be heard and her father returns after months of absence with new clothes for the festival, only to find that Durga is no more. Opu's final departure from the village is in the month of Vaisakh, just after the Charak festival. This chapter winds up all the loose threads of Opu's childhood. Aturi the witch dies on the day of Charak, ending one phase of irrational childhood terror; the music heard during the festival reminds Opu of other years when, lying securely next to his sister at night, he used to hear the same tunes; the glistening of coconut leaves in the moonlight evokes memories that tie Opu irrevocably to the place he is about to leave; Opu discovers the lost *sindur*-box his sister had been accused of stealing. In this chapter his whole childhood is gathered up and given back to him, as it were.

Parallel to this spatial and temporal inclusiveness (the past and the present coalesce; Nishchindipur remains a living presence even when Opu lives in Banaras or Calcutta) the protagonist's growth can also be seen in terms of the inclusiveness of the expanding world of his imagination, enriched by the books he reads and the stories he listens to.

Bibhutibhushan takes great care to mention in detail the titles of the books Opu reads: old Bengali novels in a neighbour's forgotten cupboard—*The Lotus and the Princess, The Bandit's Daughter*; Oliver Lodge's *Pioneers of Science*; Sir William Ramsay's *The Gases of the Atmosphere*; accounts of Cortez's conquest of Mexico and Pizarro's adventures in Peru.[8] But this is a process of enlargement and expansion of a world whose centre is always Nishchindipur. Also, side by side with his readings, his imagination is fed by oral narration of stories ranging from classical myths and local legends to folk recitals of incidents of rural interest. He forges a continuous relationship between his lived life and the experience of imaginative literature. There is no dichotomy in this world. The oceans and continents he reads about in geography books get absorbed in his desire to be a sailor, and the story he reads about the sunken Spanish ship off the coast of Porto Plata becomes part of a lifelong dream. At the end of *Aparajito*, when he is about to embark on his voyage towards South America, he tells himself that he is at last going to find that sunken treasure, reminding us of the continuity of his dreams and his ability to relate dream and reality, the written word to actual experience. The blue-eyed girl named Joan in the province of Lorraine who saved France, Grace Darling, Zuleika, the sugar fields of Martinique, Columbus' adventures—all become as much part of the boy's imaginative awareness of the world as the cruel story of Karna's death, or Romeshchandra Dutt's novels about Shivaji's truimph and Rajput valour.

In this aspect of *Pather Panchali*—the education of Opu—a very significant pattern reveals itself when we note the impact of colonial education without cultural disjunction. Where it does not alienate, it can be wonderfully inclusive. The education of Opu can be contrasted with the education of Mohan Biswas in Naipaul's *A House for Mr Biswas* (1961), which is again a bildungsroman tracing the growth of another poor brahman boy in another British colony at about the same point of time.[9] Both the boys receive their education in an eclectic fashion, reading whatever books are available, and if quite surprisingly they seem to have read some of the same books, this may be attributed to the uniformity in the pattern of education in British colonies. However, the result is quite different in the two cases. Because Opu is securely rooted he has no problem in reaching out to another culture without losing his own moorings, whereas Mr Biswas belongs to an uprooted social milieu (of Indians in the West Indies), and education for him becomes a process of alienation from his life. Mohan Biswas's own life is unsuccessful but he lives vicariously through his son, and his son's final departure to England and gradual snapping of all bonds with Trinidad seems the logical culmination of this process of estrangement. Opu too leaves India at the end of *Aparajito*, but there is something in the circular movement of the novel which assures that he will inevitably return. And while he is gone his son Kajal is left in Nishchindipur, the village of his childhood: through Kajal a new cycle continues.

The story appears to make linear advance as the family quits the village and arrives in the city of Banaras. But very soon the seamless web of Opu's world is resumed as the past begins to get absorbed into a new present. The pre-Opu chapters of the novel are a necessary link with an even older past, where the genealogy of the Roy family is traced, emphasizing continuity. We learn that Opu's father Harihar Roy and Harihar's father both lacked practical wisdom and were poets and dreamers like Opu. In the old aunt Indir's reverie past and present blended freely and she could weep over the death of a boy who passed away fifty years ago as if the sorrow was a recent one. Among the ruins of the Bengal Indigo concern at the outskirts of the village there was one tombstone which still stood unbroken:

Here lies Edwin Lermor
The Only Son of John and Mrs Lermor
Born May 13, 1853, Died April 27, 1860

The dead child had passed into history but the wild *sondal* tree which shaded the grave still showered its abundant yellow flowers on it. Opu's interest in this grave of a child who was the same age as him links the past and present in an immutable flow. On a hot summer afternoon when Opu is forced to stay indoors he explores the old wooden chests, the wicker baskets, the huge utensils under the bed whose musty smell makes him aware of a time when he was not in this world but when these things had been present. On the top shelf of their cupboard Opu finds stacks of old manuscripts written by his grandfather whose crumbling pages contain a mystery just outside young Opu's reach. This sense of mystery and wonder that Opu never loses gives the novel a timelessness that cannot be found in novels defined along temporal lines alone. After his sister's death Opu leaves Nishchindipur; his father dies in Banaras, but the central motif of the novel continues to be that of cyclic regeneration which is captured in a Sanskrit verse repeated more than once:

kale varshatu parjanyam paithivi shashya-shalini
lokaah santu niramayah
(If the rains come in time and the earth is green again, the well-being of the people is renewed.)

In Banaras Opu's father, while reciting puranic stories on the bank of the Ganga, always ended his recital with this verse. Opu used to listen to it casually as he played on the river side. Sung in the *purvi* raga these lines become the leitmotif of the Banaras section. When Opu's father dies, after the cremation and the rites are over and Opu takes the ritual dip in the cold river, this is the verse that keeps echoing in his mind. Overwhelmed by the day's events and shivering in his wet clothes, Opu imagines that he can hear his father's voice somewhere, reciting the familiar song of benediction, obliterating the image of the defeated man and perpetuating the memory of a man affirming the renewal of life. Not completion, but continuity is the essence of the novel—and this verse highlights it in the third section.

This third section of the novel ends when Opu reaches the nadir of humiliation in a crowded house in the city where his mother served as a cook. The slavery and enclosed space of this section are counterpoints to the wild freedom of Nishchindipur. In a flood of nostalgia Opu remembers all the details of Nischindipur, specially because return seems impossible now. He thinks of the magical afternoons by the forest, the yellow bird, the lemon tree, and finds that the sorrow of separation is unbearable.

The movement from village to city and the consequent nostalgia for a lost paradise which *Pather Panchali* deals with only incidentally, has turned out in the next half century to be one of the major themes in novels written in Indian languages. The historical situation of Bengal where partition of the country in 1947 resulted in the loss of a large part of the province's rural area has been specially responsible for the recurrence of this theme in Bengali novels. Nishchindipur in this context has almost become a symbolic embodiment of the lost country of childhood. *Pather Panchali* gains a retrospective significance for having been the most lyrical evocation of this theme of the wrench of man from nature. Yet Nischindipur is not merely a symbol, it is also a realistically presented village with a specific landscape and particularized vegetation. Although realism is not the basic mode of the novel, every leaf has been minutely etched. With the vivid magnifying perception of childhood Opu and Durga recognize each of the different wild berries in the village by taste and every beetle by its name. The banyan trees are also given proper names and the cows and calves are called by identifying marks. This rich specificity is somewhat attenuated in the English translation, where quite a few of these local names (of cactii, of water weeds and mango trees) are left out. These details add up to create a densely textured world of childhood in the midst of which stand Opu and Durga. They grow up in close emotional attachment but their responses to life are markedly different—Opu constantly moving outwards both in his imagination as well as in actuality while Durga remains tethered to the concrete objects of Nishchindipur. In a very characteristic episode we find the brother and sister collecting chestnuts together yet living in two different worlds. Standing in knee-deep water Durga urges Opu to hold on to her sari so that she does not slide into deeper water. At that moment

A yellow bird perched on the top of the *moynakanta* tree swayed the leaves and whistled an unfamiliar tune. Opu looked up and asked:
   'What is that bird, didi?'
   'Forget the bird Opu, hold on to the sari, tight. Don't let me slip?'[10]

Opu is forever being distracted by things beyond his reach, losing concentration on the immediate object because of flights of fancy. After listening to his mother reading out from the *Mahabharata* about the slaying of Karna, even an ordinary sunset makes him feel unbearably sad—

Then as the day wore on Shorbojoya returned to her own housework and Opu went out and stood on the verandah staring at the distant banyan tree. Sometimes the high branches were hazy in the shimmering heat of midsummer; at other times they glowed red in the still light of the evening sun. More than anything else, it was the sight of the tree stained with red colours of the evening that filled his mind with grief. In the far distance, beyond the banyan, where the sky bent down to the earth, he could see Karna, his hands labouring to drag the chariot wheels clear of the mud. Every day he laboured, every day, Karna, the mighty hero, the object of a pity which could never end. It was Arjun who won the kingdom; it was Arjun who won the fame, it was Arjun who slew his hapless foe with a bolt loosed from his charriot; but he was not the victor. Karna was the victor ...[11]

Durga has no use for these day dreams. Her concerns are earthbound and she experiences life basically through sensory perceptions, most often through touch and taste.

Opu and Durga can be regarded as two aspects of man's relationship with nature: Opu soaring high, his imagination taking him far from where his feet are planted while he continues to derive strength from the grass and weeds of Nishchindipur, the village being the centre of the concentric arcs that will chart his flight; Durga remaining attached to the palpable aspects of reality and responding to it directly through her senses. She collects shiny seeds of wild fruit for their smooth feel, green mangoes for their sharp taste. She hoards bits of a broken mirror, beads from a string, a gold sindur-box (this, in fact, she steals), quite indiscriminately and hides her treasures in a broken toy-chest. This toy-chest serves a key-function in the novel. When halfway through the story Durga's mother throws away the toy-chest and its contents in a fit of anger, it is like a ritual ending of Durga's childhood.

In the society which this novel depicts marriage is a girl's initiation into adulthood. The passing of Durga's childhood coincides with strange premonitions of departure. She begins to look at familiar sights—the bamboo grove behind the house, the road under the *gab* tree, the ferry ghat of the river— somewhat sadly. At this point in the novel the reader, like Durga, tends to connect this feeling to the inevitable separation from Nishchindipur through marriage. Durga's own thoughts of marriage, combining both anticipation and apprehension, refer to an archetypal feeling—an amalgam of joy with sadness—celebrated in marriage songs in different parts of India. In Bengal, where until the early decades of the twentieth century a girl used to be married off before she attained puberty, marriage figures largely even in nursery rhymes and childrens' songs. This separation-through-marriage motif is so insistent in Bengali culture that the entire ritual of Durga Puja, the annual autumn festival central to the life of Bengal, is regarded as the enactment of a folk myth where the goddess Durga as a married daughter comes home to her parents for a few days, and the joy of these few days is invariably tinged with the sadness associated with her impending departure. When at the end of these days the clay image of the goddess is immersed in the river, the sad music reflects the mood of the occasion. Durga is made of the earth, water and air of the village and it seems inherent in the logic of the novel that she too has to be immersed in the waters of Nishchindipur before the novel can move forward. The Bengali poet Sudhindranath Datta (1901–60) once suggested an ecological link between Bengal's climate and culture; the fact that the people of Bengal did not build stone temples and, instead, 'worshipped clay images which, once the day was done, went back to rest in the beds of superabundant rivers and ubiquitous marshes' could be related to the wet climate and alluvial soil of Bengal.[12] Durga, a barefoot, tangle-haired waif who in a way embodies the spirit of the place, also goes back to its elements when her day is over.

Marriage as a rite of passage from childhood to adulthood is pre-empted by Durga's death. Her death on a stormy night perhaps carries an unconscious echo of a central metaphor in Vaishnav poetry where Radha thinks of death as union with Krishna. Edwin Gerow has observed that Durga's death is inevitable because 'a free spirit cannot "grow up", cannot accept or come to terms with her concrete social character, which is the prison of adulthood'.[13] Although Durga has been conditioned from childhood to perform the rites for obtaining a good husband, although like all girls of her age and time she has been taught to look forward to marriage as the glittering finale towards which everything moves, her first reaction on seeing a cart carrying a new bride to her husband's home is one of terror:

'If I go away from here leaving my parents and Opu behind, will I ever be allowed to come back?' She could not conceive that she would have to leave this garden for ever, this grove of *basak* flowers, this copper-coloured cow, the shades of her favourite jackfruit tree, the smell of dry leaves and the path to the river.[14]

For her the thought of marriage is an intimation of loss and separation. In the design of the novel it seems inevitable that Durga's marriage will be pre-empted because, as Edwin Gerow suggests, her wild spirit cannot be trapped by the fetters of adulthood without losing its essence, and also because she embodies the spirit of Nishchindipur and thus cannot be separated from this setting. Had she lived longer the fetters as well as the separation would have become inevitable. As Opu grows up his horizon—both physical and mental—widens while Durga's world closes in. As a girl there are more and more things she is forbidden to do. In this case the untamed spirit that society attempts to curb is almost identifiable with the unkempt overgrown bushes and swamps of Nishchindipur—or with Vishalakshi Debi, the neglected local diety of the village. No one in the village paid much attention to Vishalakshi; her temple was uncared for, the goddess nearly forgotten—yet she was part of Opu's daydreams. When Durga was alive she was similarly abused and ill-treated by most people in the village. Yet she is Opu's most vital connection with the sensory world of nature from which he is gradually separated as he grows up. In this indirect identification between Durga and Vishalakshi Debi a process of mythicization may be seen at work.

The railroad and the train that appear as recurrent motifs in the novel also serve to define the difference between the roles of brother and sister. During their first abortive attempt to see the railway track it is Durga who takes the initiative. Under the pretext of looking for the lost calf of the red cow, the two children run in the direction of the railway but end up losing their way. Durga never manages to reach the railway which, in her and Opu's childish minds, defines the boundary of the familiar world. Later Opu, on his way to a neighbouring village with his father, sees with amazement the railway embankment and the telegraph lines, but he has not yet seen the railway train. When in between her spells of high fever Durga asks Opu, 'Will you take me to see the train one day?', already the leadership has passed from her to him, and Opu promises that he will. After Durga's death when Opu actually has the opportunity of travelling by train his excitement is not unmixed with regret. He sees from the train the vision of a sad girl standing alone under the *jamun* tree at the corner where the tracks curve away from the village. This to him is the final desertion because, so long as they were in Nishchindipur, even after she died Durga was present in every field and grove.

This sadness of departure and separation is one of the dominant moods of the novel and it has created problems for the translator in English. *Karuna rasa* is the most difficult rasa to render in English and when attempted runs the danger of being considered sentimental. In the English version of *Pather Panchali* the translators thus have had to guard against something that the author of the original is totally unselfconscious about, because sentimentality, even when it falls short of karuna rasa, is not necessarily an artistic disqualification in the Indian context. The English dictionary definition of sentimental as 'emotion in excess of the object that generates it' can easily be countered by Aziz's well-known quip in *A Passage to India*: 'Is emotion a sack of potatoes that you measure it so much for a pound?'

Herein lies another of the critical problems that the study of *Pather Panchali* raises. Is the impact of the novel culture-bound or does it touch a substratum of universal human experience? Or is it possible that the ideal of universality is itself a colonial mirage in which 'the world' is insidiously made to appear synonymous with Europe and North America?[15] The author of *Pather Panchali* was not among the English-educated elite in British India. As a homespun non-urban educated middle-class Bengali he had done a good deal of unplanned reading both in English and Bengali, but it could not have occurred to him to underplay emotion in order to conform to a standard of objectivity set in another culture. In this he is very much like some of the other major Indian novelists of the twentieth century—Tarashankar Bandopadhyay in Bengali, Gopinath Mohanty in Oriya and Shivaram Karanth in Kannada—whose strength lies in their lack of selfconsciousness about techniques and cultural dialectics. Whatever they imbibed from western literature they did without being uprooted from their own regional and cultural moorings. In their own languages these writers undoubtedly occupy very important places, but in English translation their works, like *Arogyaniketan*, *Paraja* or *The Whispering Earth*, have not received as much recognition as the more modernist Indian novels like *To Each His Stranger* (Hindi; 1961) or *Samskara* (Kannada; 1965) or *The Virgin Fish of Babughat* (Bengali; 1970) or *Steps in Darkness* (Hindi; 1958) or *Seven Times Seven is Forty-Three* (Marathi; 1978).

This fact raises several questions about the inter-relationship between language and culture, readership and aesthetic expectations, literary form and literary conventions. *Aparajito*, the second volume of Bibhutibhushan's novel, was never translated into any non-Indian language perhaps out of a fear that Lukacs seems to have predicted (before the novel was written) as a danger inherent in this particular type—the danger of sentimentality and disintegration of form—just as the other type to which *Don Quixote* and *Tristram Shandy* belong suffers from 'the danger of triviality, of being reduced to mere entertainment.'[16] But Bibhutibhushan and his readers in Bengali never seem to have worried about this danger.

When the film *Pather Panchali* first won acclaim abroad, many insensitive Indians worried that it was an unnecessary projection of India's prevailing poverty to the world outside. Continued admiration for the film at home as well as abroad does not seem to justify such apprehension. Those who have been induced to read the novel after seeing the film will soon notice that the novel's evocation of wonder is a much greater force than its depiction of poverty, even though the latter is very real.[17] Wonder is the dominant quality in Opu's perception of reality. The tracing of his ever-widening universe begins in the novel with his first outing with his father at the age of five. The sight of the first actual rabbit outside the alphabet book, the taste of wild berries, the mystery of the ruined house of the indigo planters, the temptation to touch the bright *alkhushi* fruit not knowing that this might result in an itch—this variety of sensations overwhelms the child eager for experience. As on the first page of James Joyce's *Portrait of the Artist As a Young Man*, Opu's first perceptions of the unfamiliar world are recounted in terms of visual, tactile, aural and other sensory details, all superimposed with a pristine sense of wonder. About a hundred pages and several years later we find Opu's father again admonishing him for being 'open mouthed' all the time, and this open-mouthed or wide-eyed excitement continues through Opu's boyhood and adolescence and persists even when he is an adult. Not all children have Opu's ability to invest the ordinary with a glowing quality of newness.

That Opu is not an archetypal child-hero but an individualized character becomes clear if *Pather Panchali* is compared
with another important Indian novel with a child protagonist:
Krishna Baldev Vaid's *Uska Bachpan* (1958; translated by the
author into English as *Steps in Darkness*, New York: 1959). The
world perceived by Beero, the central consciousness of this
novel, seems to belong almost to another planet. Beero lives in
an urban slum which gives him neither physical nor mental
space. While Opu's world constantly opens out, Beero's closes
in. The persistent motifs are of smoke and darkness, culmina-
ting in a climax of claustrophobia at the end with Beero's
attempt at strangulating himself. His world is drawn in mono-
chrome in different shades of black and grey, as against
the lush colours of grass, sky, fruits and flowers seen through
Opu's eyes in *Pather Panchali*. Quite unaware of *Pather
Panchali* (the novel was not even translated when *Uska
Bachpan* was written), Vaid presents in Beero an exact antithesis
of Opu. Opu never loses his childhood even when he grows up
whereas Beero has never fully been a child—his weariness and
premature wisdom are alternated by spells of pure innocence
which do not last long. Nishchindipur seems idyllic in its
superabundance of vegetation compared to Beero's world,
where there is nothing to soften the squalor. Beero's gaze is
fastened to the stagnant festering ditchwater and the hornet
buzzing over it, on the big fat louse about to drop from the
hair, or the peeling layers of his grandmother's quilt. What is
common to both novels in spite of the many differences is the
phenomenological apprehension of life as well as the ability to
live in a world of imagination and reality simultaneously that
both Opu and Beero share. Opu is poor but his poverty is
softened by the warmth of human attachment and by the
unlimited freedom to wander. Beero's world is circumscribed
not merely in its poverty, but in its deprivation of emotional
security. Opu is not particularly aware of his poverty. When
he has *halwa* in the house of a rich disciple of his father he
naively wonders why the halwa his mother makes never
tastes as good. Beero on the other hand is acutely conscious of
his poverty every waking moment of his life and fantasizes

about the happiness he would give his mother if he could earn a lot of money.

Steps in Darkness is a stark novel with a conscious focussing on the point of view of a slum child and his immediate perceptions. In *Pather Panchali* the central consciousness is Opu, but not consistently so. Often the omniscient narrative voice of the author takes over and sometimes there is an unaccounted for presence referred to as the God of the Road who speaks directly. There is less consciousness of craft here, more of the unselfconscious oral narrator's skill. The two novels stand at opposite poles both in mood and technique even though both succeed in portraying with unerring precision the consciousness of a child.

Another novel that can be set against *Pather Panchali* both thematically and technically is Manik Bandopadhayay's (Banerji) *Putul Nacher Itikatha* (Bengali; 1936: translated into English as *The Puppet's Tale*), a major landmark in Bengali fiction. Even though the setting in both the novels is rural Bengal and there is hardly ten years' difference in their publication dates, the world-views are totally antithetical. In Manik Bandopadhyay's novel nature is indifferent to human beings to the point of near malevolence, and the cycle of seasons, instead of indicating regeneration and renewal, emphasizes seasonal disease and mortality. Shashi the doctor protagonist has lost his moorings both in terms of his place in the community and in terms of his relationship with nature. He is an example of the problematic hero alienated from his environment and confronted with the task of finding his place in the universe. Opu is never dissociated with nature, even though Naresh Guha has suggested that a distance does grow between Opu and nature. Guha sees this in terms of the recurring cycle of passionate love for, then separation from, and finally *viraha* (inconsolable yearning) for Universal Nature and childhood.[18] Opu's world, despite its material poverty, is a benign one, rich in human warmth and imaginative wealth. This lyrical mode contrasts with the existential mode of Shashi's perception in *Putul Nacher Itikatha*. As a medical doctor he has very few illusions either about physical nature or human nature. The searing starkness of his vision encounters death, stagnation and carnal desire with equally ambivalent detachment. There is no hope here of transcendence.

Manik Bandophadhyay's doctor protagonist has been compared by Nabaneeta Deb Sen with Camus' doctor protagonist in *The Plague* (the French original of which was written about ten years after *Putul Nacher Itikatha*).[19] One cannot think of anything in western literature with which *Pather Panchali* can be compared. The narrative mode cannot be labelled realistic, nor is it totally lyrical or romantic. The question of whether there has been any western influence on this work in not easy to answer. The author, like his protagonist, seems to have read a great deal of English literature unsystematically, but instead of his sensibility being altered by this, it seems to have absorbed all this reading to remain essential itself but richer. The strength of the novel is its wholly unselfconscious evocation of a world where man has not become estranged from nature, where objective reality and the subjective world of imagination can still be part of an organic whole. The choice of a child as the central consciousness is crucial to the effect sought to be created and may even be the reason why the novel has withstood the passage of time, during which it has often crossed linguistic boundaries both within India and outside.

VII
GODAN

The year Premchand died, 1936, was also the most important year in his literary career. *Godan* was published that year, as was his most memorable short story 'Kafan'. In April he presided over the first session of the All India Progressive Writers Association held in Lucknow. Premchand's sharpest indictment of the capitalistic system of values also appeared in the same year in a brief essay entitled 'Mahajani Sabhyata'.[1]

While one can find a link connecting the short story, the presidential address and the essay, *Godan* eludes a neat schematic correlation with any ideology. There is very little thirties style Marxism in the novel and, paradoxically, the readers' sympathies are aligned not with the rebellious Gobar who defies the static and hierarchical society but with his patient and fatalistic father Hori to whom change or progress is inconceivable. This is the only novel where Premchand does not prescribe a remedy for a specific social problem.

Yet in the same year, in his inaugural speech at the AIPWA conference, he had insisted on the need for socially purposeful writing. Lamenting the soporific quality that dominated the then current writing in Hindi and Urdu, he pleaded for a literature that would generate dynamism, struggle and uneasiness (*gati, sangharsha aur bechaini*). He said:

By 'progress' we refer to that condition which creates in us strength and vigour, which makes us aware of our misery, which enables us to analyse the internal and external factors that have reduced us to the present state of inertia, and which attempts to remedy them.[2]

Here Premchand lays down a programme for literature, emphasizing its diagnostic as well as prescriptive functions. Yet in his own work of the period he has left behind the simple, tendentious and problem-solving approach. Both 'Kafan' and *Godan* testify to this. 'Kafan' is the almost perfect distillation of a situation that disturbs the reader with intense but understated horror. *Godan* lacks intensity of any kind but exemplifies, inspite of its flaws, how a major work can embody a latent paradox and tension generated by the contradiction of a writer's explicit ideology and his actual representation of life.

In reading 'Kafan' one remembers Engels' advice to Margaret Harkness: 'The more the opinions of the author remain hidden, the better for the work of art.'[3] The radical questions that are raised in the story do not have very simple answers. In most of Premchand's earlier work it is the poor and the weak who are exploited by society, but in 'Kafan',[4] Ghisu and Madhav have so little expectation from life that no one can exploit them. It is they who take advantage of society by stealing, cheating and not working even when they are being paid for work. Living at a sub-human level they are outside all normal mores of social behaviour. While Madhav's wife is writhing in labour pain inside the hut, Madhav and his father sit outside, greedily devouring stolen potatoes, unwilling to go and help her because the other person might grab a larger share. Ghisu recounts in detail a feast to which he had been invited twenty years ago and Madhav listens to the vivid account of food with vicarious pleasure. The wife lies dying inside.

Premchand had earlier talked about the corrosion of human values caused by money. One of his explicit purposes in writing, he once said, was to oppose the power of money. In *Godan* he creates the character Khanna to illustrate how money can dehumanize a man. 'Kafan' shows the other side of the coin: here not money but its absence dehumanizes. When Madhav's wife dies the villagers contribute towards the cost of her cremation, indicating traces of the traditional value system of an organic community where participation is obligatory, where responsibilities are shared. But Ghisu and his son repudiate such values and use the money that was meant for the dead woman's shroud in drinking and ordering the kind of lavish feast that Ghisu had eaten twenty years ago. While in the ecstasy of eating, they bless the woman who through her death has made it possible for them to gratify their gastronomic desires. As they get progressively more drunk, they become more and more sentimental about the dead woman and end with a frenzy of song and dance in her memory.

Chekhov, in a famous letter to Suvorin, says 'You are right in demanding that the artist should take a conscious (moral) attitude to his work, but you confuse two conceptions—*the solution of a question* and *the correct setting of a question.* The latter alone is obligatory for an artist.'[5] 'Kafan' is among the most memorable of Premchand's stories because, although nothing is solved here, the question has been stated precisely and disconcertingly. In a society where even hard work and honest labour do not raise a man's life very much above the animal level, how does one condemn Ghisu's and his son's rebellion? 'What kind of custom is it when a living person does not get a rag to cover his body, the dead must have a new shroud?' asks Ghisu. But even when one rationalizes that mourning is a luxury a starving man cannot afford, it is difficult to condone the action of Madhav and Ghisu. At the same time it is not easy to take the side of a society that reluctantly proffers a few coins for the dead, more to keep up a pretence than out of any real concern for human beings. The story disturbs because one sees the precariousness of our cherished human values, with bestiality on one side and hypocrisy on the other. The concentrated horror of the story arises out of the reader's inability to take sides. Since Premchand's style is totally devoid of poetry and resonance, his optimum effect can be achieved in the direction of starkness, and nowhere is this better illustrated than in 'Kafan'.

Godan does not have anything of the understatement and starkness of 'Kafan', but the mode of narration in a novel has to be necessarily different from that of a short story, specially when the novel is one to which the epithet 'epic' has often been appended.[6] Although *Godan* is a novel of suffering whose tragic end is prefigured as early as the second page, when set against 'Kafan' its total effect is benign. The basic

human values are never questioned and there is no ambivalence about ethical and moral issues, which are set forth with
almost fable-like clarity. If the rhetoric of prior intent goes
against the grain of actual narrative, the author does not seem
conscious of it, and a large number of readers and critics have
been able to read the novel without being troubled by this
paradox.

I

Godan continues to be Premchand's most widely read and
discussed novel and, after nearly half a century, according to
an opinion poll conducted by the Hindi weekly *Saptahik
Hindusthan*, it topped the list of readers' choice of novels even
in 1980. Although it has been translated into English twice,[7]
there has hardly been any critical discussion of the novel
in English. Critics in Hindi have tended to see it as a saga of
Indian rural life whose protagonist Hori is the archetypal
Indian peasant in his meekness and humanity. What is not
emphasized is that the story of the village Belari takes up just
over half the entire novel (nineteen out of thirty-four chapters
to be precise, and roughly 153 pages out of the 288 pages
of the English version); the rest delineates the uneasy interaction among certain representative characters of the urban
middle and affluent clssses—and Premchand uses a distinctly
different narrative mode to depict this segment of society.

The two strands have been simply termed 'urban' and 'rural'
by Hindi critics who emphasize the thematic aspect, and
they have either justified the necessity of the double strand or
apologized for the dichotomy. The presence of the urban
strand has been vindicated on the ground that the transition
from the feudal to the capitalistic form could not be rendered
by concentrating singly on the village or the city. Others think
that the two parts do not coalesce and Premchand is not justified in introducing the urban part.[8] But all critics agree that
the rural chapters are more authentic and artistically convincing. Critical discussion of *Godan* has generally been at
the thematic level, and the obvious dichotomy in the narrative
mode has not been commented upon. As far as the 'story' is

concerned the two strands are interlinked through contrasts and parallels, but in the 'discourse' the gap between them presents an aesthetic problem.[9]

The narrative strands at the story level are more than two and it may be helpful to number them for further reference:

A. Hori's story, and around him the story of his family and his village.
B. Gobar's story, moving from the village to the city.
C. Rai Saheb's story, straddling the village and the city.
D. Mehta and Malti's story, and that of their coterie in the city.

A and B begin together although B branches off later; C and D are closely connected because although Rai Saheb's land is in the village he has social as well as financial dealings with the urban professional class. Premchand takes care to interlink the different strands in as many ways as possible. A and C touch because there is a feudal relationship between Rai Saheb and Hori; B and D are connected because Gobar is first employed by Mirza Khurshed in the city, later by Dr Malti Kaul. A and D get linked when Mehta and Malti come to the village to help the people of Belari. Only B and C never meet or converge.

As the novel proceeds from one strand to another, the discourse pattern changes perceptibly. The change can be described as a movement from the mimetic to the diagetic mode, or from the representational to the illustrative.[10] Whatever be the terms used, the differences can be clearly seen by analysing two sample chapters—Chapter One and Chapter Seven, One concentrating on A and Seven covering C and D.

Beginning *in medias res* the first chapter conveys swiftly through dialogue and concrete details of daily work all that the reader needs to know about Hori and Dhania: their domestic situation, their age and appearance, their social level and their economic condition. Even the difference in their temperaments is brought out within the first couple of pages through Hori's statement, 'When your neck is being trampled under the tyrant's heel, the safest course is to keep on tickling his feet', (p. 5) and Dhania's counter observation, 'why so much obsequiousness for a life which did not provide even the daily

bread?' (p. 6). The author loses no time in indicating a continuity in the miserable present, an equally miserable past (three
of their six children have died without medical treatment) and a
uniformly bleak future, a glimpse of which can freeze their jest
about old age into a vision of fear. Not only does the chapter
achieve an exposition of character and setting, it sets new events
into motion, and before the chapter is over we know of Hori's
attempts to buy a cow and his dubious deal with Bhola. At the
story level, thus, the existents are established and a chain of
action initiated. (Hori is already in debt, but is willing to get into
further debt in the hope of possessing a cow). At the discourse
level, the emphasis in this chapter is on showing rather than
telling, although the direct presentation is sometimes mediated
by the author unobtrusively. The cow, the controlling motif
of the novel, is allowed to appear casually and its special
emotional significance in the cultural context emerges not from
any authorial comment but directly from Hori's imagery as he
looks wistfully at the animal which might one day become his:
'The spotted cow was walking majestically, swaying her head
gently, and flicking the flies with her tail. She looked like a
queen in the midst of maid servants.' (p. 12). The novelist's
intention is to create a credible situation and a few convincing
characters who are interesting not as vehicles of ethical values
but as individuals. The details of daily life add up to a concreteness of texture (for example reference is made to the five
items that make up the full outfit of a farmer: a staff, a quilt,
a turban, a tobacco pouch and a pair of shoes, although we
never see Hori in this complete regalia) and, with a Defoe-like
precision, Premchand sets down the detailed account of Hori's
unpaid debt and the accumulating interest. Despite a few
general comments on peasant life, on the whole the narrative mode is dramatic and objective. This continues for five
chapters and with Chapter Six, which introduces the *C* and
the *D* strands, the pattern of discourse begins to undergo a
change, this change further intensifying in Chapter Seven.

In Chapter Six a new set of characters are introduced who
have come to the village from Lucknow as the landlord's guests.
While Hori, Dhania, Bhola and the other rural characters

gradually become individualized through their action and speech, the urban characters are introduced to us directly by the narrator one by one as if they are static exhibits whose traits can be summed up once for all:

The other in high heels was Miss Malti, vivacious, lavishly made up, witty, forward, well-up in the ways of men, fond of the good things of life, cocquettish; in short, very modern. She had returned from England after qualifying in medicine and had set up private practice, counting Zamindar families among her clientele. (p. 44)

Mirza Khurshed: a fair man, brown moustached, blue-eyed, completely bald, in *achkan* and *churidar*; a devout Muslim, he had been to the Haj twice. But he drank like a fish, wore a sola topee, became very active at election time and voted nationalist ... (p. 49)

The storyteller's voice has taken over, causing a serious shift in the discourse pattern. In Chapter Seven the novel passes on from the parameter of realism which had been established in the early part of the novel to that of a fable or exemplum where the events merely serve to prove an ethical point. The moral values shared by the storyteller and the reader in real life then assume a greater importance than the values generated within the fictional world.

Chapter Seven, describing a *shikar* trip undertaken by Rai Saheb's guests, really consists of three separate exempla, all of them extolling what the author considers authentic human values and ridiculing pretension, articificiality and cupidity. The moral quality is indicated fairly early when we are told (p. 56) how each person in the party came with intentions other than the avowed one of hunting—except Mehta, and to an extent Mirza. It is also possible to see the motif of shikar as an allegorical device to highlight a predatory quality by which one creature tricks, exploits, dominates or destroys the other. It is doubtful if the author intended this consciously because he, unlike most modern readers, looks upon shikar as a positive and manly sport which tests a man's real worth. The three separate exempla relate to the three groups into which the party divides itself at the beginning of the day.

(i) *Malti and Mehta*: Most obviously the predator, Malti tries her seductive wiles on the professor of philosophy who would not at first succumb. But while crossing the mountain stream Mehta has to carry Malti in his arms and the he-man image aroused in the process causes such an euphoria that his defences crumble. Malti of course is a stock character : the bad westernized woman, a stereotype which persists in Indian culture as the vamp of the commercial Hindi film. The type occurs elsewhere too, for example in Premchand's lachrymose short story 'Miss Padma' written in 1933, and in a poem by Sumitranandan Pant[11] where the so-called modern woman is compared by turns to a butterfly, a bird, a cat, a wave and a flower, for the poet then to come to the conclusion that whatever she may be she can never have the qualities of a true woman. It is not easy to ascertain whether this stereotype owes anything to actual social reality in terms of the impact of western education and culture in UP in the early decades of the century, or whether the device of caricature is used because the westernized woman, however rare, posed a threat to the concept of womanhood sanctioned by patriarchy—which had to be reaffirmed through contrast.

When Malti and Mehta journey into the forest a tribal woman is introduced as if to underline the point of the parable. Her health and vigour, her spontaneity, her hospitality, eagerness to help others—a whole range of positive qualities are displayed in order to expose the shallowness of Malti as a woman. Malti is selfish and petulant, she wilts in the sun and gets a headache. The entire exemplum is summed up in one sentence: 'One fresh as a wild flower in full bloom, the other weak and small like a potted plant'. (p. 61). What the storyteller considers 'genuine values' is clearly sifted from artificial ones. If the tribal girl offers a female ideal by highlighting physical and emotional qualities, the male ideal as embodied in Mehta has to have an additional dimension of intellect to supplement his prowess. Having seen him as a scholar of philosophy in the earlier chapter we are now given a glimpse of his splendid muscular body. The background of the mountain, the forest and the stream complete the picture to reaffirm the organic and elemental relationship between man and nature. Malti stands outside this relationship—stiff and brittle—invoking the storyteller's contempt.

(ii) *Rai Saheb and Khanna*: These two who wander off together do not present a fable in such clear terms of polarity because neither is meant to be a paragon of virtue like Mehta. Between the two, however, the Rai Saheb is intended to come off as a more genuine person. Khanna is a banker and an industrialist, an embodiment of the values that Premchand so emphatically attacks in his essay 'Mahajani Sabhyata'. His behaviour is a fictional illustration of Premchand's indictment:

In this civilization based on capitalism, money is the sole motivation for work ... Each individual has become a predator, and the entire society his quarry. He stands outside society, interested only in seeing how much he can get out of it, how he can fool or trick society and reap the benefit ...[12]

Khanna is never far removed from his mercenary pursuits. In the forest he buys medicinal roots from a rustic vendor with the intention of selling them at a profit to gullible customers in the city. He has no taste for shikar; he has come merely to transact some financial matters with his host, yet he is meticulously dressed in a hunting outfit, an item of information meant to underline the gap between his appearance and his intentions. Unaffected by the beauty of the environment Khanna talks of business constantly and persuades the Rai Saheb to speculate in the Stock Exchange. He stands for all the negative values of a society based on the cash nexus, including a lack of physical strength and courage. His talk of non-violence is merely a cover for his cowardice and does not carry ideological conviction. True to the parable-like quality of the episode (as distinct from the realistic texture of Chapter One) a leopard suddenly appears, exposing Khanna's complete lack of manliness. The Rai Saheb is partially redeemed because he attempts to kill the leopard. Although in Chapter Two Rai Saheb has been seen as a ruthless exploiter of his bonded labourers, in this chapter Premchand emphasizes his humane side, specially to set him in contrast with the Machiavellian Khanna. Rai Saheb's sanctimonious statement, 'I think it is the reward of my past deeds that some faint glimmerings of conscience still stir me to serve my country and my fellow men', is meant to be accepted without any sarcasm on the part of the reader, although his portrait in Chapter Two is richly underlaid with irony. In this chapter the contrast between the Rai Saheb and Khanna is intended to highlight the two systems that Premchand refers to in the opening paragraph of 'Mahajani Sabhyata'—feudalism and capitalism.[13] The feudal lord has some positive human qualities while the capitalist, totally devoid of all feelings, is motivated mechanically by economic considerations. The Rai Saheb and Khanna in the shikar episode represent the two systems with unequivocal simplicity.

(iii) *Tankha and Mirza Khurshed*: In the third group the opposition is again between the genuine and the fake, the straightforward and the devious, presented in a simple diagetic mode. Tankha, we are directly told, is a mercenary opportunist—'slippery as an eel'—while Mirza is impulsive, generous and indifferent to profit and loss. If shikar is a touchstone of manliness, Mirza scores over all the other characters by actually killing a deer (Mehta shoots only a bird, the Rai Saheb tries to kill a deer as well as a leopard but does not succeed) and the fact that he can carry the dead animal for some distance makes him a superior character in the author's scale of values, superior certainly to Tankha who lacks physical stamina altogether. Moreover, Mirza has the generosity to give away the deer to a woodcutter and to pay for a feast for the whole village. That he can unselfconsciously sing and drink with the rustic crowd is another high-scoring quality in Premchand's moral tabulation, marking Mirza superior to the city slicker Tankha, who, like Malti, feels ill at ease in the village.

All three parables have a nearly pastoral intent: idealization of spontaneity and castigation of calculation; glorification of simple rustic life (it is important to see how this allegorical pastoral life differs from the mimetic rural life of the actual village Belari); admiration of physical vigour. There is nothing of Premchand's avowed social realism here in the schematic juxtaposition of persons and manipulation of events to show up moral values.

II

With the end of the seventh chapter we return from the realm of the moral fable to that of a realistic novel, from the idyllic images of rural life to the actual Belari where the rains fail to come on time, where moneylenders tighten their nooses around their victims' necks and the landlord prevents the peasant from sowing until the arrears of rent are paid. The *A* strand (Hori's story) has a continuity and logical progression, while *C* and *D* (the story of Rai Saheb and his urban friends) tend to disintegrate into separate exempla, each one driving home a palpable moral point. The Pathan episode, Mirza's *kabaddi* match, Mehta's lecture at the Women's League and the fire at Khanna's factory—these are all isolated events, not parts of a larger pattern. The fire that destroys Khanna's factory also burns Khanna's pride and goads him into confessing his past misdeeds. This is the moment of triumph for the good woman Govindi who expostulates: 'Why are you disheartened? Money is the root of every evil, the destroyer of the soul. After all what did we gain by being rich?' It seems to be of little consequence to the author that Khanna later returns to his vocation of ruthless money-making so long as the event projects Govindi as the ideal of womanhood–she being set up in binary opposition to Malti who negates the values of 'service, sacrifice and purity'.[14]

The *D* strand is clearly two-toned. Govindi is unmistakably white[15] and Malti is made to change from black to white somewhat abruptly. To convince the reader of Malti's transformation she is taken to Belari where she tells the village women how to keep their houses clean and their children healthy. Belari, when evoked in the *A* strand, has a complex reality, but when the *D* strand converges here, the village becomes a simple pastoral retreat where the men perform wrestling bouts for Mehta's entertainment and the women congregate around Malti to exclaim in admiration: 'You are wonderful, deviji, you are wonderful.' (p. 247). In Premchand's moral code the ability to become integrated with the rural folk is a test of genuineness, and once Malti passes this test the reader is expected to be satisfied with her change of heart.

Premchand is not alone among Indian writers in this nostalgic yearning for an integration of urban and rural values. Colonial education effectively cut people off from the traditional agrarian and rural ethos, creating a nostalgia for the

lost world. The attempt to re-establish this link became a central
concern for the Indian intellectual, and a far more complex
handling of this problem can be found in a novel written nearly
thirty years before *Godan*. In Rabindranath Tagore's *Gora*
(1907) the protagonist's quest is towards an integrated
wholeness. Gora wants to transcend his fragmented urban elitist
existence to become one with the larger reality of India. Gora's
aggressive religious fervour can be seen as part of his attempt to
connect himself with the faith that embraces multitudes of men,
with a tradition that goes back centuries. When he analyses his
orthodoxy in order to explain his stand to Sucharita, who is a
member of the Brahmosamaj, he has to admit:

Whether I have faith in idols or not I can't exactly say, but I respect the
faith of my country. The worship which the whole country has evolved
after so many centuries is something which I can never regard, as the
Christian missionaries do, with bitter looks.[16]

Religion is thus an anti-alienation device for Gora, an unconscious strategy of uniting himself with the masses of India, a
need which came to be emphasized in Gandhian thought later.
Gora is torn within himself as he realizes the impossibility of
his task, and this tension has continued to haunt the Indian
mind in some form or the other throughout our century. In the
last quarter of the twentieth century a writer like U.R. Ananthamurthy dwells at length on the gap between the essentially
mythic and metaphoric imagination of the Indian peasant,
and the Indian writer who is shaped by western modes of
perception.[17] Neither liberal humanism, scientific positivism,
nor Marxism will enable him to understand how the peasant's
mind works. Yet the problem is not a simple Forsterian one of
'only connect' because Ananthamurthy wonders if the 'authentic Indian peasant ... who relates to nature organically' is
himself not 'an imported cult figure of the western radicals
who are reacting against their materialistic civilization.'[18]

Gora's visits to the village also arouse in him a kind of self-
doubt about rural nostalgia. Gora is a problematic individual
whose conflict is not with society but within himself, and if his
two sojourns in the rural areas are set side by side with
the two similar sojourns of Malti and Mehta to the village, the

two-dimensional nature of Premchand's characters becomes very obvious.

Gora's visit to the village Char-Ghose-para early in the novel had been one turning point in his life, landing him in jail and making it clear to the reader, though not to Gora, that his well-meaning intervention can cause more harm than good to the oppressed villagers. By the time Gora returns to the village towards the end of the novel he has undergone a crisis of faith. He has begun to see that the traditional values that he had idealized all his life will not bear close scrutiny in their actual context. Religion, which should be a sustaining faith, is in reality reduced to oppressive rituals and the idea of a cohesive community is merely a myth. He returns to Calcutta with a greater knowledge of reality as well as the extent of his self-delusion. Unlike Malti and Mehta who are enthusiastically greeted by the villagers when they come on their altruistic mission, Gora is looked at with suspicion by the villagers. There is enough evidence of Premchand's disillusioned and clinical understanding of the rural reality in the *A* and *B* story strands of *Godan*, but the *C* and *D* strands seem to deliberately avoid realism and attempt to turn the village into an allegorical pastoral setting—a touchstone of authenticity. The poverty merely serves to make Malti 'ashamed of her gold wrist-watch, her zari-embroidered silk sari, the thick coat of powder on her face'. (p. 248).

Thus Malti and Mehta's two excursions to the village remain mere steps in the schema. On the first trip Malti spurns the tribal girl's coarse millet-bread, but during the second she sits amidst the filth and squalor and communicates with the rural women. The point of the exemplum is finally made when Mehta catches a glimpse of Malti sitting in the crowd with a dirty infant on her lap. Malti's transformation has brought her close to Premchand's ideal of womanhood, complete even in the madonna aspect.

The non-mimetic pattern of discourse is further emphasized by the way Mehta is shown building a raft in a matter of minutes (something which took Odysseus four days; but then Odysseus' raft had to cross the ocean, Mehta's only a stream) so that he and Malti can go for a boatride on the moonlit stream. Some embarrassed commentators have sought to explain away this improbability in terms of the influence of the Bombay film world on Premchand. It is well-known that he spent one frustrated year there, writing film scripts to pay off the debt incurred in connection with his journal *Hans*, but in fact this episode needs no special explanation. It fits right into the parabolic mode employed in the *C* and *D* strands, where verisimilitude is not crucial, where leopards and canoes can be produced at will if the fable demands their presence for a moral point to be made.

III

If Hori's story (strand *A*) is presented in the mimetic mode and Malti and Mehta's story (strand *D*) is entirely non-mimetic, Gobar's story (strand *B*) falls somewhere in between. The movement of Gobar from Belari to Lucknow is symptomatic of a larger pattern of transition in Indian society. The shift from the rural to the urban and the tension arising out of the consequent readjustment of values as well as the re-definition of self is a recurrent theme in modern Indian literature. Gobar's story is a schematic treatment of this theme, straddling the precipitous border between the illustrative and the representative.

When Gobar is first introduced he is a realistically conceived individual, rebellious against the system of exploitation and angry at his father's acquiescence to authority. He makes his father unhappy by criticizing the landlord's luxurious life. Hori argues:

> 'So you think there is no difference between him and us?'
> 'None. God has made us all equal.'
> 'That is not true, son. It is God who creates the high and the low. One comes into wealth after a lot of penance. Its the fruit of the deeds of our past life. We sowed nothing and we have nothing to reap.'
> (p. 18)

This is a predictable argument between two generations—one fatalistic, the other progressive. A similar argument is to be found in Mulk Raj Anand's Lalu trilogy (in the first volume, *The*

Village, 1939) and in the works of other writers connected with the Progressive Writers' Association. When Gobar departs from the village after leaving a pregnant girl of another caste behind, he repudiates what are considered the three essential elements of India's traditional social structure: caste, village community and family system.[19]

Gobar's contempt for the village community is apparent when he returns from Lucknow the first time with money in his pocket. Finding that his father has been ruined in his efforts to avoid ostracism from the community, Gobar blatantly insults all the elders in the village. He pays off some of his father's debts but also shouts at Hori for being a coward. Individual enterprise brings economic rewards, but it also entails an undermining of human values.[20] The break up of Hori's family is inevitable after this because the security and warmth of a joint family necessitates a surrender of the self which is now impossible for the newly individualized Gobar.

Spatial mobility has sometimes been regarded as a vital feature of the transition from the traditional to the individualistic pattern of life.[21] When Gobar leaves the village he dreams of Lucknow as a place of unlimited opportunity, a contrast to the stratified stasis of Belari. He manages to earn more in the city than he ever did by working on the land. Yet Premchand emphasizes not the promise of the city but its ruthless competitive aspect. When Gobar returns to Lucknow after a brief spell at home he finds that his place in the market area has been usurped by another hawker and the customers do not remember him. The contrast with Belari, where Gobar would be remembered by people no matter how long he stayed away, is obvious. The priest and the moneylender would stop Gobar on the streets of Belari to ask him about his life in the city and to find out how much he earned. If this is an invasion of privacy, it goes with caring and concern, a sense of participation that is an essential feature of village life.

The total indifference with which Gobar and Jhunia are greeted in the city causes a disorientation in them as well as a strain in their relationship with each other. In analysing the process of change from a traditional to an individualistic pattern of life in another context, an African writer, J.S. Mbiti, says that the newly individualized person 'simply discovers the existence of his individualism, but does not know of what it consists. He has no language with which to perceive its nature and its destiny.'[22] This applies to the plight of Gobar and Jhunia, of whom the latter is even less equipped to handle isolation than Gobar. A woman in a village leads a life where her role is clearly defined by the family and the community, and she always defines herself in relation to others. 'What joy can the bride have in marriage', Dhania once says, 'unless the father-in-law's house swarms with relatives?' (p. 278). In Lucknow, separated from the matrix of family and community, Jhunia is a lost character.

Chapter Twenty-Seven, describing Gobar and Dhania's stay in Lucknow (strand *B*), begins to depart from the mimetic mode. The focus shifts to the process of dehumanization that the city and the factory can inflict on the workers and Gobar becomes an illustrative figure. Labouring all day at the mill he is reduced to an automaton, unable to feel even the loss of his dead child. His love for Jhunia, wrenched from the protective network of the family and the community, turns into selfish lust. The situation seems to be a diagrammatic illustration of the Marxist concept of the alienation of the worker as stated in the *Economic and Political Manifesto*.[23] It has never been ascertained whether Premchand ever actually read Marx; his diagnosis of the human malady inherent in a certain kind of displacement and disorietation may well be based on a perception of life around him rather than on any ideological tenet. His contemporary Saratchandra also draws similar conclusions from his acute observation of human behaviour. In *Srikanta III*, there is the casual description of a colony of railroad construction workers who live on trucks. The migrant labourers wrenched from the moorings of traditional society have lost the values that made them human. Srikanta arrives there in the midst of a cholera epidemic:

In a few days I learnt how under the pretext of civilization, the rich man's greed for money can turn people into heartless animals. I was left alone with the victims of cholera in the blistering heat of the sun with only a tarpaulin sheet to give us shade. There was no one even to fetch a cup of water for the little boy who was suffering miserably. The digging of gravel had to go on; this was government work and the workers would be paid their wages at the end of the week. But the boy belonged to their community, their caste! I have never seen such behaviour in a village. The fact that these people have been uprooted from all the natural bonds of their community and the family to be put on these trucks for the sole purpose of digging gravel from sunrise to sunset, has withered all their human qualities. They are concerned only with the digging of gravel and the wages that follow. The civilized men have realized one important fact: unless you reduce human beings to the level of animals, you can not extract an animal's labour out of them.[24]

Both Saratchandra and Premchand are making observations about human behaviour that are perceptive and valid, but these observations have also to be judged in their fictional context. *Srikanta* is of course a picaresque narrative where the entire structure is episodic and linear with very little organic form, and Srikanta's encounter with the contruction labourers is just one of the many experiences that constitutes the hero's journey through life. But *Godan* is meant to be a well-knit novel with the various strands carefully interwoven. The novel loses its continuity when Gobar becomes a representative figure. We are told: 'In the village he had to put in equally hard work, but he had never felt tired. In the open fields, under the sky, even if he felt tired after the day's work, his mind remained feather-light'. (p. 225). Obviously Gobar is not being treated as an individual any more, only as a typical labourer. This Gobar is certainly not the boy seen earlier in Belari, surly and discontented, disliking his profitless labour on the land. The authorial interpretation lacks credence if one is looking for mimetic continuity. Had this observation been part of Gobar's thoughts it could be attributed to the nostalgia which distorts facts, but the author's voice is emphasized in a subsequent sentence: 'This state of mind was not peculiar to Gobar alone; it affected all the workers'. (p. 225). We know for certain that we have passed from one level of discourse to another.

Allowing for occasional exceptions the discourse pattern in *Godan* moves from the fully mimetic in strand *A* to the totally diagetic in strand *D*. The intervening spectrum contains elements of both modes, *B* being somewhat closer to mimesis than *C* (See Table below). It is also to be noted that while strand *A* depends more on concretization of setting and dramatization of action, strand *D* is predominantly verbal, containing lengthy speeches and debates. All the urban characters are highly articulate and Lucknow, their habitat, is a shadowy backdrop that never becomes as real a place as Belari.

Table

A — Mimetic; representational.
B — Mimetic, proceeding towards diagetic;
        representational, proceeding towards illustrative.
C — diagetic, containing elements of mimetic; illustrative,
        containing elements of representational.
D — diagetic; illustrative.

The difference between the mode in *A* and in *D* is almost as wide between realism and allegory, or between a novel and a fable.

IV

This duality might have generated a creative tension if Premchand had employed it as a concious fictional strategy. In effect however it weakens a novels whose popularity rests largely on the contininuing appeal of strand *A*, where the narrative derives from a compassionate understanding of individuals who are not mere vehicles of ethical principles. Hori is the victim of social determinism but he is not the allegorical figure of a sacrificial lamb. He is dignified with free will and is responsible for his action as well as his non-action. Never sentimentalized, he is shown to be shrewd and devious at times, and yet capable of remaining humane and honest. The land is the one certain value in his life and the cow is his persistent dream. He virtually sells his daughter and undergoes the humiliation of becoming a wage earner[25] in his effort to keep the useless land from which he reaps no benefit. The cow is for him an identity-marker; it appears right at the beginning, when for a few magic days he has one in his possession. His brother poisons not only the cow but the rest of Hori's life. He is crushed by the increasing burden of debt, by the pressure of the community and his own concept of prestige (*maryad*). Hori's failure is individualized, shaped not only by circumstances but by his own character, and his fate is self-evoked. Dhania too is a character conceived in the vivid mimetic mode, sharp-tongued yet tender, compassionate and querulous by turns, in open rebellion against her oppressors yet traditional in certain basic ways. The supple raciness of Premchand's language can only be approximated in the translation but even in the English version the Hori-Dhania relationship stands out indelibly as the aesthetically satisfying centre of the novel.

Although presented casually and incidentally, Sona and Rupa belong to this successful mimetic mode, as does Gobar until he moves out of the village. Even at the moment of leaving Belari, when he hides behind a tree to make sure that the pregnant Dhunia is taken in by his parents, he is a fully concretized adolescent in whom anger and fear alternate, ambition and cowardice jostle with each other. But once in Lucknow he recedes into an illustrative figure, a cardboard cutout exemplifying the urban corrosion of values.

Rai Saheb is another of the characters who oscillates between two modes. He depends almost entirely on his estate for survival and is not above squeezing his poor tenants when he needs money, yet the author is eager to show that he is really not a crook at all—perhaps merely a victim of circumstances. Premchand makes some effort to suggest that the Rai Saheb and Hori, poles apart in the economic scale, are really two facets of the same social system. The same false sense of prestige that compels Hori to bribe the police officer and spend money for his daughter's wedding makes the Rai Saheb contest an election or fight a court case when neither can afford such actions. Both virtually sell their daughters and are alienated from their defiant sons. Exploitation is the central problem of the novel. Despite his facade of charity the Rai Saheb exploits his tenants, and he in turn is exploited by the capitalist financier and by the government rules, and as a result he himself sinks upto his neck in debt. The opposing elements in Rai Saheb's character do not necessarily make him a complex character—they merely betray the wavering of the author's sympathy. To Premchand Rai Saheb's feudalism with its old world graces is preferable to the mercantile values of Khanna; but when seen in relation to his tenants in Belari his charm and culture ring hollow.

The coexistence of contradictory traits—that make Hori and Dhania believable human beings—in Rai Saheb's case merely create an uncertainty of effect. The uncertainty persists in all the urban characters and most of all in Mehta who sometimes appears as an idealized figure, sometimes as a satirically presented character whose theory and practice do not match: but never as a convincing living human being. Usually the unconvincing quality of the urban strand is explained by Premchand's greater understanding and preference for rural society. But in reviewing Gordon Rodarmel's translation of the novel, Gayatri Spivak makes an interesting point about the uncertainity of norms and the uneasiness of values actually existing in the urban North Indian society which Premchand was attempting to describe: 'The adolescent verbosity, the moral irresponsibility, the sentimental yet ruthless relationship between the sexes, are documentarily true not only of the lugubrious haute bourgeoise of the author's own day, but apply to that class (as a class somewhat less powerful now than then) in Northern India today.'[26] This statement is elusive and unverifiable but it contains an interesting suggestion nevertheless, making us take a fresh look at Mehta's improbable contradictions (for example his reactionary ideal of womanhood and his interest in a woman's gymnasium, or his concept of selfless love and the ferocious possessiveness of a lion) and relating them to the contradictions of the feudal and traditional ethos which Premchand inherited, an ethos which was superficially being modified by western liberal, colonial and industrial values.

It has been observed earlier in this chapter that in *Godan* the rhetoric of prior intent often goes against the grain of actual narrative. This happens at more than one level. The acquiescent serf Hori becomes the sympathetic centre of the novel and Gobar's protest proves ineffective and jejune. Premchand's professed mode of 'idealistic realism'[27] does not find a vehicle here; the fable-like anterior mode which he thought he had discarded comes back to punctuate the realistic narrative. Failure of the rural community becomes a more convincing theme than either the condemnation of urban life or the nostalgic evocation of the pastoral world.[28] The feudal land-lord turns out to be a pathetic victim of circumstances, more worthy of sympathy than stricture.

These are not necessarily weaknesses. In a time of quick transition—both in life and in literary modes—major works of literature often appear that are marked by such internal contrariety. *Krishnakanter Will* and *Anna Karenina* can be seen as examples of this. Lukacs and Goldmann have shown how there may be a fruitful contradiction between an author's explicit ideology and the actual representation of life in his work. Tolstoy gives 'plenty of incorrect and reactionary answers'[29] to social questions, yet that does not detract from his value as a social realist. Engels put forward this idea with reference to Goethe and Balzac. The paradoxes inherent in *Godan*, though not always creative or fruitful, have to be seen in terms of the peculiar tensions of Premchand's time, both in his life and in Indian literary history.

VIII
SAMSKARA

Soon after being published in Kannada in 1965 *Samskara* by U.R. Anantha Murthy became the centre of a controversy which was somewhat non-literary in nature.[1] The novelist was accused of attacking brahmanism, and in 1970 when the novel was made into a film as well as later when the English translation was serialized in the *Illustrated Weekly of India*. the controversy was raked up again and again. The memory of that debate has made many subsequent readers of *Samskara* respond to it as an essentially realistic novel dealing with a social problem. When the English translation was published as a book in 1976 the translator, A.K. Ramanujan, took some pains to correct this view and provide the right perspective to approach the novel. He called it 'an allegory rich in realistic detail' in the Translator's Note, and further explained in the Afterword: 'As in an early Bergman film, the characters are frankly allegorical, but the setting is realistic. An abstract human theme is reincarnated in just enough particulars of a space, a time, a society'. (pp. 144–5). Ramanujan also offers a brief anthropological interpretation of the novel which is so perceptive that most subsequent discussion of *Samskara* has been dominated by Van Gennep's *Rites of Passage*, except perhaps V. S. Naipaul's. Naipaul's oft-quoted views on *Samskara*[2] emphasize the obscurantism of the 'barbaric civilization' where readers accept the 'premises of the novel that are so difficult for an outsider: caste, pollution, the idea of the Karma-given self, the anguish at the loss of caste-identity'.[3]

The significance of *Samskara* as a modern Indian novel lies precisely in the author's attempt to exploit the tension between two world views. The ethos which Naipaul found incomprehensible—where identity is determined by *karma* and *varna* —is shown in collision with a new awareness of self, partly conditioned by existential thinking. The difficult and uneasy process of transition between the fixed settled order of life and the still inchoate stirrings of self is part of the thematic concern of the novel. Although largely allegorical in texture and mythic in its conscious structure, the novel does not repudiate the demands of realism. Thus both in content and form it can serve as an illustration of the kind of mutation that a western form has undergone in India.[4]

The novel begins with an emphasis on the static quality of life as lived in the brahman colony of Durvasapura village. The word 'routine' is repeated three times on the first page, high-lighting the lack of spontaneity in Praneshacharya's acts: 'he offered food and flowers to the gods *as he did everyday*' and whether he recited sacred texts or worshipped Maruti or bathed his invalid wife, it was all part of an 'unfailing daily routine'. This inflexible schedule is established before a ripple is allowed to form in this stagnant life by the announcement of a death.

If the first chapter begins with the daily routine of Praneshacharya, the spiritual head of the community and the protagonist of the novel, the second chapter places him in the larger routine of the village, in the annual cycle of life in the *agrahara*. There is uniformity in all the brahman houses, both in their interiors ('cockroaches in the buttermilk shelves, fat rats in the store'), and the exteriors ('sacred balsam plants in the backyard' and flowers for worship). The lives of the people are ordered by rituals and festivals and seasonal food varia-tions. There are vows to keep in each month and occasional feasts for death, marriage and initiation. 'The brahmins' lives ran smoothly in this annual cycle of appointments'. (p. 16).

The word 'smoothly' is ironic because the total effect, far from being smooth, is claustrophobic. The details emphasize the sterility of this ritual-bound existence and although the Tungabhadra river runs behind the houses, the flowing water seems to have no relation to the enclosed lives of the brahmans. Time stands still here and kinships and hostilities continue for generations. Human relationships are graded and merits measured according to a prescribed scale. A brahman cohabiting with a low-born prostitute may be condoned, but not if he has eaten food cooked by her. (pp. 5–6). A man neglecting his wife at the time of her death may be pardoned, but not if he has omitted performing the annual rites for his dead parents. All decisions are made according to an inviolable code made centuries ago.

It is tempting to contrast the stability and norm of traditional society as depicted in R.K. Narayan's novels with the world created by Anantha Murthy at the beginning of *Samskara*. Almost all of Narayan's novels present a stable world temporarily threatened by disruption yet resilient enough to survive the crisis and return to normalcy. On the whole Malgudi has a positive life-affirming and abiding quality. Anantha Murthy on the other hand intends to depict a decadent structure which, once jolted out of its groove, cannot be reintegrated again. The arena of conflict turns out to be not the village but the mind of Praneshacharya, the most conscious point of the community, and it is not possible for him to return to the original stasis once the age old routine is questioned. The negative quality of the society is emphasized by the fact that no birth or marriage takes place during the course of the novel; the crisis is precipitated by an epidemic and almost the entire novel is stalked by death. The marriages that are briefly mentioned are all joyless and sterile.

At the beginning of the novel Praneshacharya is very much a part of this decadent agrahara, although he is not corrupt, avaricious or hypocritical like the rest of its inhabitants. But halfway through the novel the shell of custom and ritual breaks and he can no longer fit into a stratified and coded existence. Thereafter Pranesh has to search out a role for himself; there is no safe social niche he can occupy and his actions and speech can no longer be determined by the expectations of society. In a sense the issue here is diametrically the opposite of that in Raja Rao's *The Serpent and the Rope*. The identity of Raja Rao's hero has to be determined by his brahmanhood. Praneshacharya had also, without conscious thought, always equated the brahmanic code with his essential self. Suddenly he is confronted with a question: if he rejects his brahmanhood what remains of his self? When he sets out on his journey in the third section of the novel he is really in search of the residual self that remains after the outer shells are discarded.

The question need not necessarily be connected with Hinduism. It is the universal problem of a man who has equated himself with a particular role for so long that the role becomes his self, and without the role he feels lost. One remembers Shakespeare's Richard II after his abdication in a similar crisis of identity:

> ... I have no name, no title
> No, not that name was given me at the font,
> But 'tis usurp'd: alack the heavy day!
> That I have worn so many winters out,
> And know not now what name to call myself.
> ....
> Let it command a mirror hither straight,
> That it may show me what a face I have,
> Since it is bankrupt of his majesty.
>
> (*Richard II*, IV. i. 255–9 and 265–7).

In a gesture of self-dramatization Richard looks for his face in a mirror and breaks it. For Richard the tragedy is unequivocal because the loss of kingship reduces him to nothingness. But for the protagonist of *Samskara* the loss of his social role is in a way a liberation and a regeneration, as was the loss of caste for Gora more than half a century before him.

To read *Samskara* as a critique of orthodox Hinduism is to limit it severely. It is a novel worth taking notice of not simply because it repudiates a decadent value system but because it is a novel where the physical and the metaphysical fuse; where the interiority of an individual's social predicament is dealt with in its psychological complexity; and where the problem—although uniquely personal—reflects also the crisis of a civilization in which through a painful process a collective code is giving way to individual choice. More than half a century ago Lukacs had made a distinction between the epic hero and the problematic hero of the novel. The epic hero, according to him is, strictly speaking, 'never an individual', and the epic is always about the destiny of a community, not an individual:

the hero of the novel is the product of estrangement from the outside world. When the world is internally homogeneous, men do not differ qualitatively from one another; there are of course heroes and villains, pious men and criminals, but even the greatest hero is only a head taller than the mass of its fellows, and the wise man's dignified words are heard even by the most foolish. The autonomous life of interiority is possible and necessary only when the distinctions between men have an unbridgeable chasm; when the gods are silent ...[5]

Praneshacharya's passage through the novel is from being one kind of hero to another. In the beginning he is one of a homogeneous community albeit a head taller than the others, connected by indissoluble threads to the community whose fate is crystallized in his own. By the end of the novel he is a lonely man unsupported by the community or God, and he has to chart out his own path.

II

'All commentary', said Northrop Frye, 'or the relating of the events of a narrative to conceptual terminology is in one sense allegorical interpretation.'[6] *Samskara* seems specially open to this mode of interpretation because not only the events but the objects described in the novel seem to be invested with allegorical overtones.

The flowers that bloomed in the gardens of the brahmans in Durbasapura village are a good example of this. These were used only for purposes of worship and were never enjoyed for their beauty or fragrance. The description in the beginning of Chapter Two is a remarkable palimpsest of realistic details and allegorical nuances, setting the tone for the rest of the novel. Only the flowers in Naranappa's yard were different because these were solely meant for 'Chandri's hair and a vase in the bedroom'—for sensuous human enjoyment and not for divine consecration. The basic polarity of the novel—between direct involvement in the sensuous aspects of life and a detachment through the denial of the senses—is thus indicated as early as p. 15.

Much later in the novel when Praneshacharya's accustomed world is shattered by Chandri's touch, he becomes aware of this opposition between his earlier detachment and present involvement:

> He had so far not desired any of the beauty he had read about in the classics. All earthly fragrance was like the flowers that go only to adorn the god's hair. All female beauty was the beauty of Goddess Lakshmi ... all sexual enjoyment was Krishna's when he stole the bathing cow-girls garments and left them naked in the water. Now he wanted for himself a share of all that. (p. 77)

The opposition between the values of the agrahara which the Acharya shares in the beginning of the novel and those of the renegade Naranappa extend beyond the framework of religion or custom. With allegorical clarity two fundamentally different responses to experience are presented: one emphasizing order and restraint, the other abandonment and passion. The smell of the night-queen-bush in Naranappa's yard broadcasts to the village the latter message: 'In the darkness of the night the bush was thickly clustered with flowers, invading the night like some raging lust, pouring forth its nocturnal fragrance. The agrahara writhed in its hold as in the grip of a magic serpent-binding spell.' (p. 15). The words 'darkness', 'night', serpent', 'magic', 'lust' and 'writhe' combine to evoke an irresistibly erotic aura which is a threat to the life-denying values of the agrahara.

The snake has obvious sexual connotations, and by transference so do the snake-like braids of women. 'Chandri wore her black snake-like hair in a knot' (p. 15) and Padmavati's 'snake-braid coming down her shoulder, over her breast' (p. 123) unsettles the Acharya's equilibrium. Even when unbraided, women's hair has erotic associations. At Sripati's call Belli came out, 'her hair washed in warm water, wearing only a piece below her waist, naked above, waves of hair pouring over her back and face'. (p. 40). The hair—serpent—eroticism thread runs throughout the novel, for example in the description of the woman acrobat at the fair as 'serpentine ...

all curves', and of Belli 'like a snake writhing in the sand'. All the brahman women in contrast to Chandri, Belli, Padmavati and the heroines of legend and myth are depicted as frigid with 'dwarfish braids' and withered bodies. The fisherwoman in Ravi Varma's painting and the Muslim wife of Jagannatha the poet excite the imagination of the starved brahmans.

In this novel the allegorical implications may dominate over the realistic details, but one comes across similar images in many other Indian novels with presumably realistic intent: outcaste or lower-caste women are often endowed with a greater sexual vitality than their high-born counterparts. While analysing two novels by Tarashankar Bandopadhyay, *Ganadevata* (Bengali; 1942) and *Panchagram* (Bengali; 1943), Rajat K. Ray has attributed this difference to certain socio-economic factors.[7] Notions of purity and pollution are more strongly embedded as one goes higher up the social scale. The taboos of touch practised over a long period of time tend to curb spontaneity. Chastity, satitva, and penance for widows cannot be the values of a class that does not have the economic means to enforce them, and apparently this freedom adds to the uninhibited naturalness of the lower-caste women. But the formation of a stereotype is not a simple process. The easy availability of lower-caste women may also have imbued them with a greater erotic aura in the male imagination.

In *Samskara* women are made to carry the allegorical burden while the male protagonist is invested with subjectivity and agency. Praneshacharya's invalid wife epitomizes the diseased sterility of the entire agrahara. The life principle embodied in women has dried up in the rigidity of the orthodox community, while outside this enclosed world there is a celebration of life made more desirable by contrast. Sripati's wife would tighten and twine up her thighs when he approached her, but there was always Belli in the outcaste hutments: 'her body . . . the colour of the earth, fertile, ready for seed, warmed by an early sun' (p. 37). The sensuousness of the women outside the agrahara is raised to a symbolic level by repeated mythic references to Urvashi, Menaka and Matsagandha–'temptresses of the sages'. The apsaras stand outside social and ethical parameters and embody in them the feminine essence unfettered by familial relationships. Thus the withered Bhagirathi and the luscious Chandri are both symbolic figures in the dream landscape of Praneshacharya's journey.

If the serpent connotes feminine eroticism the tiger is repeatedly associated with masculine lust. Naranappa's flowers invade the night like some 'raging' lust and the word 'raging' connects it to his own lust when he confronts Chandri's body 'like a raging striped tiger'. (p. 45). Praneshacharya, reflecting on his encounter with Chandri in the forest, recognizes 'his body's tigerish lust' (p. 81) which lay dormant all this time under pity and compassion. Once having tasted blood 'now the tamed tiger is leaping out, baring its teeth'. (p. 82). Also by implication the tiger gets associated with other aspects of life that fall outside the rarefied and attenuated brahmanic existence—the world of violent entertainment and crude joy. The acharya is horrified by the 'tigerish world of cock-fights' which threatens his new-found values as well as his orthodoxy.

It is not difficult to see that the serpent and the tiger belong to the Dionysian world which constantly threatens the repressed orderliness of the brahman agrahara. And yet to call this force evil, as R.K. Kaul does, may lead one to overlook its complexity. According to Kaul 'the alternative to brahmin austerity is a surrender to darkness and the demons'.[8] Although the world represented by Naranappa's drunken debauchery is certainly not meant to be a positive whole, one should hesitate to apply the words 'evil' or 'darkness' to them in any direct or unambiguous way. It is the obverse of the barren life of Praneshacharya, full of privation and sacrifice, where all spontaneity is stifled and where 'God has become ... a set of tables learned by rote'. (p. 92). Both Naranappa and the Acharya represent distortions of certain values—restraint, control and denial in one and abandonment to the senses in the other. They are two sides of the same coin, hence their names are constantly coupled by the author, making them adversaries in an almost equal combat.

The agrahara of Durbasapura is famous, we are told at the outset, because of the learned man Praneshacharya ('the crest jewel of Vedanta') *and* because of the notoriety of its scoundrel Naranappa. The basic issue of combat is also set forth fairly early: 'The real challenge was to test which would finally win the agrahara: his own penance and faith in ancient ways, or Naranappa's demoniac ways.' They seem to be trying to reach the same destination in their different ways. The Hindu puranas do refer to two ways of attaining god—one as a devotee and the other as an enemy. Like Jaya and Vijaya, the legendary brothers, Naranappa's way is a negative one. [9] In this duel the Acharya is constantly on the defensive. To win over Naranappa is for him an egotistical need. Words like 'desire' and 'lust' in this connection (p. 22) reveal the vulnerable spots in his armour of self-control.

Naranappa's death, instead of being his defeat, turns out to be a victory. Like the dead body in Ionesco's play *Amédeé* or *How to Get Rid of It*, the corpse of Naranappa seems to swell gradually and fill up the whole agrahara in a metaphorical as well as real sense. The plague, the stench, the panic the confusion, everything seems to proclaim the power of the dead man. The security of the place gives way to a reign of fear—there are vultures during the day, ghosts at night: 'Naranappa's challenge was growing, growing enormous like God Trivikrama who started out as a dwarf and ended up measuring the cosmos with his giant feet'. (p. 34).

It is Pranesh's egotistical and wilful belief that he can bring Naranappa back to the right path by his own faith and austerities, perhaps another face of the same exaggerated greed for virtue which made him wed an invalid and make a penance out of daily life. When he wakes up in Chandri's arms in the forest he knows he has lost the battle: 'I was defeated, defeated—fell flat on my face' (p. 100), and the reader knows that the basis of the defeat goes back to his early life when he stifled the natural instincts of a man in his zeal of piety. This defeat at least breaks the shell within which he has lived all these years. But even in Pranesh's awareness of a new freedom Naranappa remains an unattainable ideal. He thinks with envy and admiration: 'How fearlessly Naranappa lived with Chandri in the heart of the agrahara.' Suffering the pain of being born into a new life all over again, the Acharya remembers his friend Mahabala and adversary Naranappa, both of whom had vanquished him: 'Naranappa, did you go though this agony? Mahabala, did you go through it?' (p. 112).

Chandri, who links the two combatants, is a liminal creature.[10] By virtue of her profession she is both outside structured society as well as recognized by it. Like the river Tunga she is in the village but unshackled by it. 'How can sin ever defile a running river? It is good for a drink when a man is thirsty, its good for a wash when a man is filthy, and it is good for washing god's images with. It says yes to everything, never a no.' (p. 44). Chandri is a symbol rather than a realistic character embodying a natural wholeness and an instinctive spontaneity which Praneshacharya can never achieve. After their union in the forest they cross the river back together. He is stricken by his conscience but Chandri is totally untroubled: 'She was a natural in pleasure, unaccustomed to self-reproach'. (p. 68). She eats most naturally when others in the agrahara are supposed to be fasting because she knows that no rules apply to her. She handles the crisis of the corpse in a way which is the exact antithesis of the behaviour of the brahmans. She acts spontaneously and calls some people in who unceremoniously cremate the body. Thus while controversy rages in the agrahara about technical points in the' *shastras*, the corpse quietly disappears.

Chandri offers her body to Praneshacharya in a spontaneous gesture and in this act of compassionate giving paradoxically becomes a mother figure. The word 'compassion' so far associated in the novel with brahmanical virtues, specially with the Acharya's feeling for his sick wife, is now suddenly transferred to Chandri. She gives him food, she gives him solace and by giving him access to her body brings grace to his life. It is allegorically significant that this act, which opens out a new world of naturalness and wholeness to the Acharya, happens in the forest—outside the frame of stratified society. His vision suddenly becomes clear, as if a veil which has for all these years separated him from the throbbing pulsating world, has dropped. All his five senses are awakened int a sudden joyous awareness.

Below were green grass *smells*, wet earth, the wild Vishnukranti with its sky blue flowers and the country Sarsaparilla, and the smell of a woman's body sweat … He gazed, he listened till his eyes were filled with the *sights*, his ears with the *sounds* all around him, a formation of fireflies. 'Chandri', he said, *touched* her belly and sat up. (p. 67; italics added.)

When the Acharya begins to see for the first time, his experience is not an unadulterated celebration of life; it has a disturbing and disconcerting aspect as well. Not only does he perceive beauty, even ugliness gains a sharper focus. When he returns from the forest to bathe his ailing wife as usual he notices for the first time with horror her 'sunken breasts, her bulbous nose, her short narrow braid', and recoils.

From this point onwards the novel progresses more in the mind of Praneshacharya than in terms of physical action in the outer world. Sitting with his dying wife in the lonely agrahara with the vultures circling the sky, amidst the overpowering stench, the raucous cries of crows, in hunger and heat, he reviews his entire life in the light of the new situation. The sacrifice in marrying an invalid, he can now see, was in the nature of an investment, a strategy to gain credit in the spiritual world. The confidence that he was basically 'a man of goodness' was merely hubris. From being a dispassionate observer of life he wants to be an involved participator, and in the process he becomes human. No longer the paragon of virtue, he has the desire to 'tell lies, to hide things, to think of one's own welfare', and yet he wants to live openly and fearlessly like Naranappa. Although he cannot resolve his contradictions he becomes aware of them.

During his introspection a Sanskrit verse about man's sinfulness comes to his mind.[11] He rejects it because he does not think what he has done is sinful. Readymade words and verses will not do for him any more, yet his desire to free himself from self-deception conflicts with his inability to tell the truth to his community. He does not know whether his inhibition comes from his own pride or a reluctance to let down a community which trusts him, but he knows that his intense self absorption will not be easily purged.

At least one of his realizations is unambiguous. He discovers the simple and unselfconscious response to life: 'Just sitting coolly under a tree he had become a fulfilment, a value, to be, just to be, to be; keen, in the heat, the cool, the grass, the green, the flower, the pang, the heat, the shade, putting aside both desire and value'. (p. 83). The act of love, instead of being an initiation into adulthood, paradoxically reveals to him the sensory nature of a child's consciousness. He is aware of the sensuous joy of swimming in cold water and rolling in the sun-warmed sand afterwards, caressing the neck of a playful calf as its hair began 'raising in pleasure' and it 'began to lick his ears and cheeks with its warm rough-textured tongue'. All these become vividly rich experiences. His senses become more acute; the smell of grass and wet earth hit his nostrils and the stars become as sharply visible as to a child's eye. After making love to Chandri he feels as if he has 'fallen into his childhood, lying in his mother's lap and finding rest there after great fatigue'. (p. 67).

There is a direct contrast between the prematurely old Praneshacharya, burdened with the wisdom of all the scriptures in his head and the responsibility of the moribund agrahara on his shoulders—and the child freshly responding to creation. The place of his rebirth is outside the arena of stasis: the agrahara, where time is stagnant and space is enclosed. The death stench of the agrahara and the smell of wet earth and grass in the forest are allegorically juxtaposed. A similar contrast between the repressive human world and the lush wholeness of nature can be seen in a novel of another century from a different culture. Arthur Dimmesdale in Hawthorne's *The Scarlet Letter* (1850)—another priest-protagonist grown prematurely old while carrying the heavy burden of erudition and respect—becomes human only when he is in the forest with Hester. Nature seems to proffer consecration to their union. But the priest is the repository of the collective values of a community; hence his violation of these values appears to have momentous significance. Their predicaments are strangely parallel. Both Praneshacharya and Dimmesdale undergo agonizing conflicts within themselves, but they emerge in different directions. Hawthorne not only wrote of a Puritan society, he himself held an uneasy identification with some of

its values; but the Acharya cast off his old life and was born
to a new one outside its bounds, while the body of his adversary
rotted in the village.

III

In myth and literature a journey often serves a symbolic
function, embodying a transition between two modes of existence, one which is over and the other yet to begin. During a
journey a man is outside his professional and commercial
framework, freed from an expected pattern of behaviour
and therefore able to contemplate the true nature of his self,
unfettered by social and familial role playing. From Odysseus'
nine-year-long journey home to Sinbad's voyages, from
Christian's progress from the City of Destruction to the Celes-
tial City, *Gullivers Travels*, *Gil Blas* and *Don Quixote*—
consciously or unconsciously journeys have been used as
different kinds of fictional strategy. They are a quest for home, a
quest for treasure, for salvation, for experience, and not
infrequently a quest for one's own true self.

'During a journey, when you shed your past, your history,
the world sees you as just one more brahmin'—realizes
Praneshacharya after he leaves his village, and his continuing
education seems to consist in learning to adjust to the world's
image of himself divested of his fame and reputation. In the eyes
of the villager he meets on the way he is no longer the 'crest
jewel of Vedanta' but a lowly brahman perhaps on his alms-
collection round, and if he loses even the external appearance
of a brahman, his image of himself would have to be further
adjusted to the world's image of him as an anonymous, casteless
wanderer.

After leaving the village Pranesh's initial impulse is one of
freedom—freedom from duties and obligations to the community, from the necessity of concentration: 'O to be without
desire. Then one's life becomes receptive'. A peeling of
layers has begun. He has lost all lustre and influence and is
preparing himself to bear the loss of public esteem as well. What
remains after all the layers are peeled off will be his true self. But

till the end of the novel the doubt remains about whether a man can ever reach that core. *Samskara* as well as *The Serpent and the Rope*, both written by South Indian brahmans, have often been read in terms of the protagonist's brahmanic world view. Brahmanism in India, and especially in the south, is perhaps more than a caste distinction: it is a special mode of apprehending reality, an experience that pervades all aspects of a man's life, going beyond his conscious mind. The episode in Raja Rao's *Kanthapura* where the Gandhian protagonist Moorthy, who has repudiated caste, still trembles as he enters a pariah house and cannot drink water there, points to the unconscious and instinctive conditioning of a brahman which makes his caste indistinguishable from his self.

Praneshacharya, an orthodox brahman, wakes up to the need to affirm 'the essential and vital importance of personal identity in one's life'.[12] And in critically examining this need S. Nagarajan in a review of the novel questions why Pranesh should feel it altogether necessary to shake off his brahmanism, 'as if Brahminism were *per se* an excrescent superimposition on his personal existence'.[13] 'Is it not?' is the rather naive counter-question of the uninitiated, to whom the debased brahmanism that the Acharya has to shake off seems a stultifying and depersonalizing force, something that negates the basic values of life.

The dissatisfaction of orthodox critics with the novel has risen from the fact that the novelist's repudiation of Hinduism does not take into account the metaphysical base on which the Hindu way of life is built, but that seems an irrelevant issue as far as the novel is concerned. The central character attempts to reject a rigid dehumanizing code of religious custom as socially practised, and one does not know till the end of the novel, how far he will succeed in his attempt to liberate himself. The Acharya's struggle is with the dogma that stifles spontaneity, and it is difficult to see why the novel should at all be read as a repudiation of Hinduism. Putta too is a Hindu, though not in any reflective or philosophical way. He accepts the caste stratification implicitly, yet it has not hampered his naturalness, his zest for living, his ability to say with a great life-affirming inclusiveness, 'On the whole I like people.' (p. 112).

Putta is an important character in the third section of the novel, and perhaps the only character other than Pranesh who is presented at some length. All the other characters make very brief appearances. Naranappa is already dead when the novel begins. Whatever we know about him is through others' memories. Chandri disappears abruptly from the novel after shattering Praneshacharya's fortress of certitudes. Mahabala's story is recounted in two pages. The other brahmans matter not individually but in their choric function. Putta is the only character who stays with us for the last quarter of this brief novel.

If Naranappa is Praneshacharya's arrogant anti-self, Putta is another antithesis of the Acharya in a lower key. As against the Acharya's total self-absorption, Putta does not have very clear sense of the boundaries of his self. In an utterly unselfconscious way he is willing to involve himself in other people's lives for no reason at all and participate in everything that goes on around him. He himself describes his amorphous fluidity thus: 'Oh Putta? Our Putta. If you let him go you lose him; but find him, he'll never leave you.' (p.121). In a paradoxical way this rather common wayfarer has achieved that ideal state Pranesh has only had a glimpse of—'to be, just to be ... putting aside both desire and value'—participating in life fully and without asking questions. When Putta's riddle 'One plays, one runs, one stands and stares' is answered by the Acharya 'the fish plays, the water runs, the stone stands and stares', we notice the Acharya's inability to be either the fish or the water, conditions to which Putta effortlessly aspires. Like the stone he can merely stand and stare with disengaged eyes at the world of ordinary pleasures and the bustle of the market place.

Unlike him Putta forms relationships easily and can even wear down the Acharya's forbidding silence by asking riddle after riddle until he is forced to reply. While Praneshacharya is contemplating his philosophical predicament, Putta chatters all the time, casually offering him coconut and jaggery when he is hungry, guiding his steps and conducting his business for him when the need arises. By the end of the day Putta has even managed to evoke a fatherly love in Pranesh. (pp. 117– 18). The Acharya has never felt this tenderness before, and frightened by this feeling he withdraws immediately.

Plagued by no metaphysical dilemma and directed by no special purpose, the earthy Putta can change his plans and destination without a moment's notice. He accepts his own ambiguous caste status without any questioning or resentment. He is allowed to wear a sacred thread, but cannot join the brahmans in their community feast. Once in the fair he instantly becomes a part of it, revelling in its noise, its smells, its visual delights. He must take a look inside the bioscope box to see the God of Tirupati as well as the Bombay concubine. A natural participator, he must throw a coin to the beggar, buy a yard of ribbon for his wife, drink a bottle of soda water and belch with satisfaction, bet at a cockfight—and thereby take his full share of the festive spirit of the fair.

The description of the bustle and noise of the fair-ground with its appeal to the senses brings back distant but distinct echoes of the experience Pranesh had when the veil dropped from him in the forest and he too could live through his senses:

'*noise* of reed pipes', '*smell* of burning camphor and job sticks', '*song* of the balloon seller', the invitation to *see* the peep show ...' (p. 113; italics added)

'he *smelt* the forest smells and *smelt* the sarsaparilla anew, he *looked* at the Vishnukranti flowers as if mere looking was wealth, he *felt* the water on his skin and the fishes pricked at his ticklish toe-spaces and ribs'. (p. 84; italics added)

The difference between the two experiences is that one is outside the world of men and the other right in its midst; one brings solace to him in the cooling shades of the forest, the other in the naked glare of the sun only hurts and distracts him. The stimuli Praneshacharya responds to belong to the pure vision of childhood and make him one with nature. The child's integrated vision does not last long. Once on his way with Putta, the concreteness of his sensory perceptions recede to the background, making him take refuge in abstractions, unable to respond to Putta's simple though crude human warmth.

Are the worlds of the forest and the fair essentially different? Is the Acharya's aim merely to recapture the feeling for nature or does his integrated vision also include the unregenerate and messy human world?

The cockfight is to the Acharya the most traumatic experience of this world. Putta naturally belongs here and the Acharya can only stand and stare from a horrified distance. The noise, the dust, the colour and the smell that assault his sheltered and secluded sensibility in the market place are nothing compared to the intensity of violence at the cockfight. The 'sharp cruel looks' of the audience, the glint of knives tied to the roosters, the throaty inhuman sounds of the people en-couraging the fight, frighten him and make him reconsider his decision. He may have rejected the brahmanical world of austerities and penance, but he will not be able to embrace this demoniac world of cruelty either. He wavers, realizing the dual aspect of the newly discovered world: 'One part of lust is tenderness and the other part is a demoniac will' (p. 117)—the forest and the fair-ground being its two components. Chandri's touch has the compassion of a mother, but Padmavati's elongated dark eyes hold teror for him: 'The bird is paralysed by the stare of the black serpent'. (p. 123). The choice is no longer as simple for Praneshacharya as it was when he had looked at the stars in the forest and smelt the wet earth in its pristine innocence.

Afraid that he might be sucked into this world of betting and hard bargaining where he is quite out of his element, Praneshacharya pre-empts Putta's attempt to get him a good price for his gold ring. This is a tentative gesture of his rejection of the values of the market place, but he cannot take a decisive step. Having lost one form yet not having the courage to acquire another, Praneshacharya continues in this indeterminate stage. He is actuely conscious that the eyes of others are investing him with a form he himself does not yet possess. Padmavati's eyes looking at him from behind the threshold terrorize him because 'as soon as eye meets eye who knows what shape the unformed will take'. (p. 123). He is afraid that the decisive moment had come and is going to be different from what he has imagined it will be.

Within a space of half a page (p. 123) he passes through various stages—fear, feeling of vulnerability, then an upsurge of desire obliterating all selfconsciousness. He passes from being a victim to being the predator: 'Bird ravaging, bird ravaged, the knives.' The violence of the cockfight enters his consciousness and will not be exorcized. In a brief kaleidos-copic flash the scenes of his life come back to him, bringing images of Naranappa, Mahabala, Chandri, Bhagirathi. His inner kaleidoscopic vision is paralleled by the outer kaleidoscopic view of the market place—the medicine man, the acrobat, the mutilated bodies of lepers reminding him of the rotting corpse he has left behind. To escape their collective assault he enters the temple, a familiar sanctuary, safe from the clamour of unorganized life outside, postponing the decision yet a while.

Seated in a line with hundreds of hungry brahmans waiting for a meal Praneshacharya is assailed by three kinds of fear. First, since he is in the impure period of mourning his proximity has polluted everyone, whether they know it or not. This is a legacy of his earlier life and he cannot rid himself of the deep seated belief that his act is wrong in itself. Second, if he is found out his act will hurt a whole crowd of people and the chariot festival will have to be cancelled. Does he have the right to disturb the lives of so many others? Third, if he is found out what will people think of him? Thus he is back to square one—to self-absorption and an egotistic greed for social approval, now complicated by an uncertainty about what is right.

The brahmans immediately try to place him by asking him about his community, sub-group and descent line. If he does not confess now to the community about his change of status he will be forever plagued by the fear of discovery. And if he does, all other lives will be muddied. The pain and agony involved in his dilemma is the pain of transcending one mode of existence to go into another, from being an epic hero in Lukacs' sense of the term to being the problematic hero of a novel. He has reached the point from where he will have to take the decisive leap, but he turns back and runs away from it with unwashed hands.

In this unformed stage he even thinks nostalgically of an earlier time of certainty—the stability and order of the agrahara before the crisis—the lighting of lamps in the evening when the cows and calves return and their milk is offered to God. That security is forever lost to him. The decision to confess gets more and more difficult the longer it is postponed. Twice he comes very close to making the decisive gesture of publicly owning upto his changed self, but twice he shies away. He lacks the will to make an existential decision. His journey back to Durbasapura also happens rather than is voluntarily undertaken. The cart going in that direction has only one seat, so Putta is left behind and Praneshacharya moves towards his third climactic moment of confession.

His decision to confess is linked with the knowledge that a rotten corpse awaited his return. The absence of the corpse might still change his decision, but the novel ends before his journey is over, tracing an arc from the static certainty of 'the crest jewel of Vendanta' to the dynamic uncertainty of a lonely individual in search of a code by which he can be completely himself without any self-deception. Because of the contradictions within the protagonist the possibilities of the arc ever being completed into a circle are remote. The Acharya will perhaps never attain the pure state he is searching for but his experience has encompassed a larger area than he had known in his earlier limited life, and he cannot go back completely to his former position. The strength of the novel lies in its multidimensional tracing of the quest, doing what Chekhov in that famous letter demanded that a work of art ought to do. Between 'the solution of a question and the correct setting of a question, the latter alone is obligatory for the artist.'[14] The question that Anantha Murthy 'correctly set' in this novel has proved to be both artistically compelling and socially unsettling, even if the solution is not provided.

Appendix I

THE FIRST CHAPTER OF BANKIMCHANDRA CHATTERJI'S *INDIRA* (1873)

At last I was going to my husband's home. I was nearly nineteen but I had not yet lived with my husband's family. The reason: my father was wealthy, my father-in-law was not. Shortly after my marriage my father-in-law had sent for me, but my father did not let me go. He sent a message saying 'My son-in-law must first learn to make money, otherwise how will he maintain my daughter?' My husband was then twenty years old. Hurt by these words he took a vow that he would earn enough to support his family. There were no railroads in those days and the journey to the west was a hazardous one. He made this difficult journey on foot without any money and reached Punjab. Someone who can do that can also earn money. He made a lot of money and sent it home but did not return for seven or eight years, nor did he keep in touch with me. I used to be miserable and angry. How much money does one need? I felt annoyed with my parents for having mentioned money at all. Is money greater than my happiness? There was a lot of money in my father's house; I could play with money if I wished. I used to think that one day I would lie down on a bed of money to see how it felt. I told my mother, 'Let me spread some money one day and sleep on it.' 'Silly girl', she said, but she understood. I do not know what devices she used, but a little before the time when this story begins my husband returned to his village. There was a rumour that he had earned enormous wealth by working at the comissariat (I hope that is the word). My father-in-law wrote to my father, 'By your blessing Upendra (my husband's name is Upendra. Older readers should please forgive me for referring to him by name. In the current fashion of course it is possible to call him 'my Upendra') is now able to support his wife. I am sending a palanquin with bearers. Please send our bride to us.

Otherwise, with your permission, we will get our son married elsewhere.'

My father was amused at the ways of the new rich. The palanquin was upholstered with brocade. There was silver work on the wood and the carrying poles ended in sharks's heads made of silver. The maid who came with it wore a raw silk sari and had gold beads around her neck. Four dark and bearded men from Bhojpur accompanied the palanquin.

My father Haramohan Dutta had been rich for many generations. He smiled at the display and said, 'Indira, my daughter, it seems I can't keep you any longer. Go now, I will bring you back soon. Don't laugh at the way these people seem to have swollen up from the size of a finger to that of a banana trunk.'

I replied him silently in my mind saying, 'My heart has swollen from the size of a finger to that of a banana trunk. Don't laugh at me.'

My younger sister must have understood. She said 'Didi, when will you come again?' I squeezed her cheeks.

Kamini said, 'Do you know anything about your husband's house?'

'Yes I know', I said. 'It is the garden of paradise where the god of love shoots his flower arrows to fulfil the lives of men. When they reach this place women become apsaras and men turn into sheep. The cuckoo sings there every day, the south wind blows in winter and the moon shines even on a new moon night.'

Kamini laughed and said 'Go to hell'.

(my translation)

Appendix II

THREE EARLY STATEMENTS ON REALISM

1. From the Preface to *Indulekha* (1888). O. Chandu Menon on Realism.

Others...asked me, while I was employed on this novel, how I expected to make it a success if I described only the ordinary affairs of the modern life without introducing any element of the supernatural. My answer was this: Before the European style of oil painting began to be known and appreciated in this country, we had—painted in defiance of all possible existence—pictures of Vishnu as half man and half lion...pictures of the god Krishna with his legs twisted and turned into postures in which no biped could stand.... Such productions used to be highly thought of, and those who produced them were highly remunerated, but now they are looked upon by many with aversion. A taste has set in for pictures, whether in oil or water colours, in which shall be delineated men, beasts, and things according to their true appearance, and the closer that a picture is to nature, the greater is the honour paid to the artist. Just in the same way, if stories composed of incidents true to natural life and attractively and gracefully written, are once introduced, then by degrees the old order of books, filled with the impossible and the supernatural, will change, yielding place to the new.

Translated from the Malayalam by
W. Dumergue (Calicut, 1965 reprint), p. xiv.

2. Mirza Mohammad Hadi Ruswa in the Preface to *Zat-i-Sharif*.

It is the practice of some contemporary writers to frame a plot in order to prove a particular point and then fill in the details accordingly. I make no objection against them, but I shall not be at fault if I simply say that my method is the opposite of theirs. I aim simply at a faithful portrayal of actual happenings and am not concerned with recording the conclusions to be drawn from them . . . I have not the inventive power to portray events that happened thousands of years ago, and moreover I consider it a fault to produce a picture which tallies neither with present day conditions nor with those of the past—which, if you study the matter carefully, is what usually happens. Great ability and much labour is usually required to write a historical novel and I have neither the ability nor the leisure to do it.

Translated from the Urdu and quoted by
R. Russel in *The Novel in India: Its Birth and Development*, ed. T.W. Clark (London, 1970), pp. 132–3.

3. Mirza Mohmmad Hadi Ruswa in the Preface to his incomplete novel *Afshai Raj*.

The most paying and interesting subject of study in this world is what happens to human beings; not only their external behaviour, but also their inner feelings and thoughts. These can be depicted through a novel provided an effort is made to present the picture truthfully . . . We should not give ourselves unnecessary trouble by trying to base our novels upon the lives of persons about whom we cannot know anything in detail. In our own circle of friends and relations there are bound to be many whose experiences are truly strange and fascinating. The trouble is that we do not pay heed to them because we cannot spare time from pouring over the tomes of the histories of Alexander the Great, Mahmud of Gazni, Henry VII, Queen Anne, Napoleon Bonaparte, etc.

Translated from the Urdu and quoted by
Khushwant Singh and M.A. Husaini in the
Introduction to the English translation of *Umrao Jan Ada* (New Delhi, 1982 reprint), pp. vii–viii.

CHRONOLOGICAL LIST OF MAJOR NOVELS IN INDIAN LANGUAGES PUBLISHED BETWEEN 1801 AND 1900

1801	*Rani Ketaki Ki Kahani (Hindi)*	Insha Allah Khan
1801	*Pratapaditya Charitra (Bengali)*	Ramaram Basu
1801	*Bagh-o-Bahar (Urdu)*	Mir Aman
1803	*Nasiketopakhyan (Hindi)*	Sadal Mishra
1823	*Kolikata Kamalalaya (Bengali)*	Bhabanicharan Bandopadhyay
1825	*Nabo Babu Bilas (Bengali)*	——
1832	*Nabo Bibi Bilas (Bengali)*	——
1852	*Phulmoni-o-Karunar Bibaran*	Catherine Hannah Mullens
1857	*Yamuna Paryatan (Marathi)*	Baba Padmanji
1857	*Aitihasik Upanyas (Bengali)*	Bhudeb Mukhopadhyay
1858	*Alaler Gharer Dulal (Bengali)*	Pyarechand Mitra
1861	*Muktamala (Marathi)*	Lakshman Moreshwar Halbe
1862	*Hutom Pyanchar Naksha (Bengali)*	Kaliprasanna Singha
1865	*Durgeshnandini (Bengali)*	Bankimchandra Chatterji
1865	*Raja Madan (Marathi)*	Babaji Krishna Gokhale
1866	*Kapalkundala (Bengali)*	Bankimchandra Chatterji
1866	*Ratnaprabha (Marathi)*	Lakshman Moreshwar Halbe
1868	*Fasana-i-Azad (Urdu)*	Ratan Nath Sharshar
1868	*Manjughosha (Marathi)*	Naro Sadashiv Risbud
1868	*Karana Ghelo (Gujarati)*	Tuljashankar Mehta
1869	*Bangadhip Parajaya (Bengali)*	Pratapchandra Ghosh
1869	*Mirat-ul-Arus (Urdu)*	Nazir Ahmad

1869	*Mrinalini (Bengali)*	Bankimchandra Chatterji
1870	*Vichitrapuri (Marathi)*	Keshav Lakhsman Jorvekar
1870	*Suhasyavadana (Marathi)*	Vaman Krishna Deshmukh
1870	*Devrani Jethani ki Kahani (Hindi)*	Pandit Gauri Dutt
1871	*Mochangad (Marathi)*	Ramchandra Bhikaji Gunjikar
1871	*Bodhasudha (Marathi)*	Keshav Balwant Kelkar
1872	*Sri Rangaraja Charitra (Telugu)*	Narahari Gopalakrishnamma Shetty
1872	*Bant-ul-Nash (Urdu)*	Nazir Ahmad
1872	*Vama Shikshak (Hindi)*	Munshi Kalyan Rai
1873	*Swarnalata (Bengali)*	Taraknath Gangopadhyay
1873	*Visha Vriksha (Bengali)*	Bankimchandra Chatterji
1873	*Indira (Bengali)*	——
1874	*Banga-Vijeta (Bengali)*	Romeshchandra Dutt
1876	*Hambir Rao Ani Putlibai (Marathi)*	Vishnu Janardan Patwardhan
1876	*Deep-Nirvan (Bengali)*	Swarnakumari Debi
1877	*Chandrasekhar (Bengali)*	Bankimchandra Chatterji
1877	*Madhavi-Kankan (Bengali)*	Romeshchandra Dutt
1877	*Rajani (Bengali)*	Bankimchandra Chatterji
1877	*Bhagyavati (Hindi)*	Pandit Sraddharam Phullauri
1878	*Rajasekhara Charitra (Telugu)*	Kandukuri Veeresalingam
1878	*Maharashtra Jeevan Prabhat (Bengali)*	Romeshchandra Dutt
1879	*Rajput Jeevan Sandhya (Bengali)*	——
1879	*Narayan Rao Ani Godavari (Marathi)*	Madandev Venkatesh Rahalkar
1879	*Prathapa Mudaliar Charitram (Tamil)*	Samuel Vedanayakam Pillai

1879 *Chhinna-Mukul (Bengali)*	Swarnakumari Debi
1881 *Rajsingha (Bengali)*	Bankimchandra Chatterji
1881 *Ratnalakshmi (Gujarati)*	Jehangir Ardeshir Talyarkhan
1881 *Nihsahaya Hindu (Hindi)*	Radhakrishan Das
1882 *Anandamath (Bengali)*	Bankimchandra Chatterji
1882 *Pareeksha Guru (Hindi)*	Srinivas Dad
1883 *Satyavati Charitram (Telugu)*	Kandukuri Veeresalingam
1884 *Tabut-un-Nusuh (Urdu)*	Nazir Ahmad
1884 *Debi Chaudhurani (Bengali)*	Bankimchandra Chatterji
1884 *Kulina Ane Mudra (Gujarati)*	Jehangir Ardeshir Talyarkhan
1885 *Shyama Swapan (Hindi)*	Thakur Jaganmohan Singh
1886 *Bou-Thakuranir Haat (Bengali)*	Rabindranath Thakur (Tagore)
1886 *Sansar (Bengali)*	Romeshchandra Dutt
1886 *Sitaram (Bengali)*	Bankimchandra Chatterji
1886 *Nutan Brahmachari (Hindi)*	Balkrishna Bhatt
1887 *Saraswatichandra I (Gujarati)*	Govardhan Ram
1888 *Indulekha (Malayalam)*	O. Chandu Menon
1890 *Pan Lakshyant Kon Gheto (Marathi)*	Harinarayan Apte
1890 *Mysorecha Vagh (Marathi)*	———
1890 *Hriday Harini (Hindi)*	Kishorilal Goswami
1891 *Martanda Varma (Malayalam)*	C.V. Ramana Pillai
1891 *Chandrakanta (Hindi)*	Devkinandan Khatri
1891 *Bibasini (Oriya)*	Ramashankar Ray
1892 *Saraswatichandra II (Gujarati)*	Govardhan Ram
1892 *Kankavati (Bengali)*	Trailokyanath Mukhopadhyay
1893–5 *Kamalambal (Tamil)*	B.R. Irajam Ayyar
1893 *Nutan Charitra (Hindi)*	Ratanchandra Pleader
1893 *Flora Florinda (Urdu)*	Abdul Halim Sharar

1894	*Samaj (Bengali)*	Romeshchandra Dutt
1894	*Sukhasarvari (Hindi)*	Kishorilal Goswami
1894	*Naye Babu (Hindi)*	Gopalram Gahmari
1896	*Chandrakanta Santati (Hindi)*	Devkinandan Khatri
1896	*Jaya (Hindi)*	Kartikprasad Khatri
1897	*Sukumari (Malayalam)*	Joseph Muliyil
1898	*Saraswatichandra III (Gujarati)*	Govardhan Ram
1899	*Umrao Jan Ada (Urdu)*	Mirza Mohammad Hadi Ruswa
1900	*Saraswatichandra IV (Gujarati)*	Govardhan Ram
1900	*Monomoti (Assamese)*	Rajnikant Bordoloi

NOTES

Chapter 1.   From Purana to Nutana

1. For a recent example see editorial in *Book Review* (New Delhi), VII : 4 (January-February 1983).

2. George Steiner has stated:

    Dialectical materialism holds that literature, as all other forms of art, is an 'ideological superstructure', an edifice of the spirit built upon foundations of economic, social, and political fact. In style and content the work of art precisely reflects its material historical basis. The *Iliad* was no less conditioned by social circumstances (a feudal aristocracy splintered into small rival kingdoms) than were the novels of Dickens which so strongly reflect the economics of serialization and the growth of a new mass audience. Therefore, argues the Marxist, the progress of art is subject to laws of historical necessity. We cannot conceive of *Robinson Crusoe* prior to the rise of the mercantile ideal. In the decline of the French novel after Stendhal we observe the image of the French bourgeoisie.

    See *Language and Silence* (1958; rptd. Harmondsworth, Penguin, 1969), p. 328.

3. According to Ian Watt, Defoe's 'total subordination of the plot to the auto-biographical memoir is as defiant an assertion of the primacy of individual experience in the novel as Descartes' *cogito ergo sum* was in philosophy'. See his *The Rise of the Novel* (1957; rptd. Harmondsworth, Penguin, 1963), p. 15.

4. Northrop Frye, *Anatomy of Criticism* (1957; rptd. Princeton, 1971), p. 305.

5. Quoted by Ian Raeside, 'Early Prose Fiction in Marathi', in *The Novel in India*, ed. T.W. Clark (London, 1970), p. 90.

6. *Indulekha*, trans. W. Dumergue (Madras, 1890), p. xx. The translation was reprinted in 1965 by Matrubhumi Printing & Publishing Company, Calicut.

7. Ibid., p. xi.

8. For further discussion of this point see the following papers by Edwin Gerow: (a) 'The Quintessential Narayan', first published in *Literature East & West*, X : 1–2 (1966), reprinted in *Considerations*, ed. Meenakshi Mukherjee (New Delhi, 1977); (b) 'The Persistence of Classical Aesthetic Categories in Contemporary Indian Literature : Three Bengali Novels', in *The Literatures of India : An Introduction*, ed. Edward Dimock and others (Chicago, 1974).

9. V.S. Naipaul, *India : A Wounded Civilization* (London, 1976). See chapter entitled 'A Defect of Vision', pp. 98–116.

10. Edward Said, *Beginnings : Intentions and Method* (New York, 1975), p. 81.

11. Once, at an international conference, a Japanese critic made a point about the essentially linear construction of Japanese narrative from *The Tales of*

Genji to the modern Japanese novel, as against the orchestral construction to be found in European novels such as Tolstoy's *War and Peace*. The first protest against this observation came from the novelist Yukio Mishima (1925–70) who cited his own works as examples to the contrary. See *Expression, Communication and Experience in Literature and Language, Proceedings of XII Congress FILLM*, ed. Ronald Popperwell, pp. 19–20.

12.   'The Origin of Genres', *New Literary History*, VII (1976), p. 163.

13.   Recorded by Shantiswarup Gupta in his *Hindi tatha Marathi Upanyas ka Tulanatmak Adhayayan : 1900–1950* (New Delhi, 1976), p.4.

14.   See Kusumavati Deshpande, *Marathi Kadambari : Pahile Shatak*, Part I (1953), p. 20.

15.   *Studies in European Realism*, English trans. by Edith Bone (London, 1978), p. 244.

16.   See Anand, *Conversations in Bloomsbury* (New Delhi, 1981).

17.   Quoted by K.M. Munshi in *Gujarata and Its Literature* (Bombay, 1953).

18.   *Telugu Novel*, ed. Adapa Ramakrishna Rao and others, Vol. I (Secunderabad, 1975), p. 13.

19.   *Mirat-ul-Arus*, English translation by G.E. Ward (1896; rptd. London, 1908), p. 8.

20.   Quoted by R.E. Asher, 'The Beginnings of the Tamil Novel', in *The Novel in India*, p. 184.

21.   *Indulekha*, trans. Dumergue, p. 368.

22.   Harry Levin, *The Gates of Horn* (New York, 1963), p. 3.

23.   In support of this the following may be noted :

(a) Bhalchandra Nemade (Marathi novelist and English teacher) stated in an interview : 'In terms of reading, more influence must have come from European and English than from English ones. I dont't think we are very much attuned to British culture. I suspect this is true of most writers of my generation. Personally, the writers I like are Homer, Shakespeare, Dostoevsky and Kafka'. (Translated from Marathi by Chandrasekhar Jahagirdar). See *Vagartha/ 23* (October 1978), p. 10.

(b) In his Introduction to the Penguin anthology of Indian writing edited by him, Adil Jussawalla says, 'It is no accident that the most potent foreign influences on Indian writing today are Camus, Dostoevsky, Kafka and Sartre', and speculates on why the literature of Britain had ceased to have much meaning for the Indian writer. See *New Writing in India* (Harmondsworth, Penguin Books, 1974), p. 27.

(c) In a seminar paper he read in January 1977 in New Delhi, U.R. Anantha Murthy (the Kannada novelist and English teacher) said : 'And the names and examples that dominated our discussion were different from those fashionable ten years ago. In place of Eliot and Yeats, dear to us for the impact of Indian philosophy on them, we used now the ideas of Camus, Kafka, Sartre, and Lukacs. We admired the achievement of Russian masters, who seemed better influences for us than the Anglo-Saxon writers who are antimetaphysical and pragmatic in their outlook. Wasn't the Russian literary scene before the revolution very similar to ours, in its struggle between the Westernizers and the Slavophiles? Dostoevsky with his metaphysical brooding was closer to the Indian temperament than the writers of the novels of manners.' See his 'Search for an Identity', in *Identity and Adulthood*, ed. Sudhir Kakar (New Delhi. 1979), p. 106.

(d) In a seminar at Mysore in January 1981, Mulk Raj Anand contrasted the empirical attitude of British novelists and their concern with realism with the metaphysical and epistemological concerns of European, American and Latin American novelists and their awarenesss of other modes of perception. He proposed that it is with the latter that the Indian writer has a more natural affinity.

Chapter 2. Pilgrim Prose and the Novel of Purpose

1. The Serampore Mission Press started work in 1800; it printed books in Bengali, Marathi, Hindi and Assamese. A Tamil printing press was set up in Tranquebar in 1710. In 1836, Reverends N. Brown and O.T. Cotter set up the first printing press in Assam. Missionary presses for printing Hindi books were set up in Ludhiana in 1836, in Allahabad in 1838, and in Agra in 1840.

2. See Sukumar Sen, *History of Bengali Literature* (1960; rptd. New Delhi, 1971), p. 163. Also see *Modern India and the West*, ed. L.S.S. O' Malley (1941; reprinted London, 1968), p. 392.

3. R.S. McGregor, 'The Rise of Standard Hindi and Early Hindi Prose Fiction', in *The Novel in India* (1970), p. 148.

4. Incidentally, this might have been the beginning of the official parting of ways between Hindi and Urdu as two separate languages. The year he died Premchand spoke bitterly about the damage this separation had done to Indian life and culture:

    It was all the doing of the college at Fort William, which gave recognition to two styles of the same language as being two different languages. We cannot say whether there was some kind of politics at work even then or whether the two languages had already diverged substantially. But the hands which split our language into two also thereby split our national life into two.

    See Amrit Rai, *Premchand: A Life*, trans. Harish Trivedi (New Delhi, 1982), p. 352.

5. Mrs Collins was the daughter of Reverend Hoxworth of Travancore who extended pressure on this princely state for the abolition of bonded slave labour. She married Richard Collins who became Principal of the Christian Mission College at Kottayam. Mrs Catherine Hannah Mullens' father was Reverend Alphonso Francois of Calcutta; her husband was J. Mullens, another missionary. See in this connection Meenakshi Mukherjee, 'Mrs Mullens and Mrs Collins: Christianity's Gift to Indian Fiction', *The Journal of Commonwealth Literature*, XVI (August 1981), pp. 65–75.

6. This quotation is from a typed copy of the original English novel, made available to me by Professor K. Ayyappa Paniker of the University of Kerala, Trivandrum. All quotations from *The Slayer Slain* are from this typescript, hence page numbers have not been cited.

7. Ian Watt, *The Rise of the Novel* (1963), p. 64.

8. From the preface in English to the original Bengali volume, published in 1852.

9. Ibid.

10. Krishna Chaitanya, *History of Malayalam Literature* (New Delhi, 1971), p. 260.

11. Banerjee brought out a new edition in 1957 carrying a valuable introduction by Suniti Kumar Chatterji. The text is based on a neglected copy then available in the National Library at Calcutta, which did not seem ever to have been read during the one hundred years since its publication.

12. Saroj Bandopadhyay, *Bangla Upanyase Kalantar* (Calcutta, revised and enlarged third edition, 1976), p. 81. Bengali.

13. *Bankim Rachanavali*, I, p. 216.

14. Pandit Gauri Dutt, *Devrani Jethani ki Kahani* (Patna, 1966), p. 34.

15. See Chapter I, note no. 18.

16. Krishna Chaitanya, p. 260.

17. Quoted by A.I. Mayhew, 'The Christian Ethic and India,' in *Modern India and the West*, ed. O'Malley, p. 324.

18. See Bhalchandra Nemade, 'Marathi Kadambari: Prerana va Swaroop', *Anustuv* (September-October 1980).

19. Preface to *The Bride's Mirror*, trans. G.E. Ward (1866; rptd. London, 1908), p. 2.

20. Preface to *The Repentance of Nusooh*, (1877; rptd. London, 1884), p. 2.

21. Ibid.

22. Georg Lukacs, *Theory of the Novel* (Cambridge, Mass., 1971)), p. 103.

23. Satyajit Ray in an interview; see *The Cineaste Interview on the Art and Politics of the Cinema*, ed. Dan Georgakas and Lenny Rubinstein (Chicago, 1983), pp. 385–6.

Chapter 3. Recreating a Past: Fiction and Fantasy

1. *Myth and Reality* (1963; Harper Torchbook edition, 1968), see pp. 134–8.

2. 'Aitihasik Chitra' ('Images of History'), in *Rabindra Rachanavali*, Vol. XIII, p. 477.

3. Translated from quotation in Bijitkumar Datta, *Bangla Sahitye Aitihasik Upanyas* (Calcutta, 1963).

4. Quoted by Datta, ibid., from a contribution to *Wednesday Review* (1905).

5. See Pramila Gupta, *Premchand aur Harinarain Apte ke Upanyason ka Tulanatmak Adhyayan* (Delhi, 1970), pp. 174–5. Hindi.

6. See Lakshmi Sagar Varshneya, *Adhunik Hindi Sahitya* (1971), p. 184. Hindi.

7. See his essay on Bankimchandra in *Rabindra Rachanavali*, Vol. XIII, p. 891.

8. See G.C. Bhate, *History of Marathi Literature* (Pune, 1939). p. 148.

9. See Srikumar Bandopadhyay, *Bangasahtiye Upanyaser Dhara* (1939; rptd. Calcutta, 1956), p. 13. Bengali.

10. K.S. Chiplunkar translated the *Arabian Nights* into Marathi in instalments which appeared between 1861 and 1865. See T.W. Clark (ed.), *The Novel in India* (London, 1970), p. 80

11. Rai, *Premchand: A Life*, pp. 18–19.
12. See *Yuganta: End of an Epoch* (Delhi, 1973).
13. See Bijitkumar Datta.
14. Quoted in Bijitkumar Datta.
15. 'Vrihattara Bharat' ('Greater India'), in *Rabindra Rachanavali*, Vol. XIII, p. 351.
16. David Kopf has offered this further explanation in 'Dimensions of Literature as an Analytical Tool for the Study of Bengal, 1800–1830,' *Bengal: Literature and History*, ed. Edward Dimock (Michigan, 1967), p. 116: 'The Bengali intellectual of the early 1800's found himself insecure psychologically not because he was in the centre of a spatial encounter between two cultures, but also because he found himself centred lengthwise in a newly discovered historical dimension. The Orientalists infused him with their image of an Indian golden age, while the Serampore missionaries transmitted a Protestant's view of the medieval dark ages. Both left the Bengali with a faith in the perfectibility of all mankind. On the one hand the intelligentsia viewed itself as the representative of an exhausted culture organically disrupted by historical circumstances but capable of revitalization. It is not surprising then that the Bengalis themselves should interpret the nineteenth century heritage as a renaissance.'
17. Thus in Marathi Vasudev Govind Apte translated all the novels for his *Sampurna Bankimchandra* volumes; *Anandamath* is included in Volume II (Bombay, 1923). In the same year appeared Bhau Sridhar Kulkarni's adaptation of *Anandamath* (Pune, 1923). In Telugu, Venkata Parvateesha Kavulu translated all of Bankimchandra's novels, including *Anandamath*. Later came another translation by O.Y. Doraswamayya in 1924 and a more recent one by Kamalasanudu in 1966.
18. This was apparently started by Sri Aurobindo but the bulk of the translation was done by his brother Barindrakumar. The earliest translation was the one by Nareshchandra Sengupta and published under the title *The Abbey of Bliss* (Calcutta, 1906). Another translation, with some alteration in the latter part of the novel, was *Dawn Over India* (New York, 1941) by Basanta Coomar Roy, an expatriate Indian.
19. *The Heart of Aryavarta* (1925; rptd. 1980), p. 114; quoted by Bimanbihari Majumdar, 'The Ananda Math and Phadke', *Journal of Indian History* (April, 1966), p. 94.
20. K. Satyam, 'Three Pioneer Indian Novelists', *Indian Literature* (July-August 1981), p. 79.
21. *Bankim Rachanavali*, II, p. 336.
22. See Jogeshchandra Bagal's editorial introduction to *Bankim Rachanavali*, I, x-xii.
23. 'Bharat Kalanka' ('The Shame of India'), reproduced in *Bankim Rachanavali*, II, p. 241.
24. 'Phadke was arrested in the nizam's territory on July 2, 1879, by Major Daniel and Syed Abdul Hak, Police Commissioner of Hyderabad, on the charge of raising an army of 200 men for looting the Khed Treasury. The looting was designed with a view to equipping himself with money to raise an army for "destroying the English". At the time of the arrest he was

dressed as a Sannyasi and was known as "Kashikar Buwa", a hermit of Benares.' See Bimanbihari Majumdar, p. 94.

25. Ibid., p. 96.
26. *Bankim Rachanavali*, I, p. 670.
27. Ibid., p. 727.
28. My attention was drawn to this coincidence by Dr Sisirkumar Das, Tagore Professor of Bengali, University of Delhi.
29. Bimanbihari Majumdar, p. 98.
30. Ibid., p. 100.
31. For a detailed account of these additions and subtractions, see Bimanbihari Majumdar, pp. 100–1.
32. During 1982, the centenary year of the publication of *Anandamath*, the Left Front goverment of West Bengal withheld state patronage to the cele-brations on the ground that the novel propagates communal intolerance. A heated controversy ensued in the Bengali press.
33. See the epilogue to *Rajsingha*, *Bankim Rachanavali*, I, pp. 661–2.
34. This notwithstanding the fact that some Muslim characters have been por-trayed sympathetically: for example Mubarak and Dalani Begum in *Raj-singha*, Ayesha in *Durgeshnandini*.
35. *Bankim Rachanavali*, II, p. 305.
36. In his paper 'Communal consciousness in late nineteenth century Hindi literature' (presented at a seminar on Social Transformation and the Crea-tive Imagination held at the Nehru Memorial Museum and Library in Feb. 1983) Sudhir Chandra has shown that communalism and nationalism coe-xisted in varying degrees in most of the Hindi writers of the period, particu-larly Bharatendu Harishchandra (1835–85), Pratapnarain Mishra (1856–94) and Radhacharan Goswami (1859–1923).
37. In the essay 'Bangadesher Krishak' he pleaded for equitable distribution of property among the people in order to achieve true progress in the coun-try: 'Then, instead of five or six gentlemen who speak softly at a meeting of the Indo-British Association, we would hear the great ocean-roar of six crore peasants.' (*Bankim Rachanavali*, II, p. 314).
38. See Sukumar Sen, *History of Bengali Literature*, p. 214.
39. 'We catch a glimpse from the Lieutenant's report of a female dacoit by name of Devi Chaudhurani, also in league with (Bhawani) Pathak. She lived in boats, had a large force of barkandazes in her pay, and committed dacoities on her own account, besides receiving a share of the booty obtained by Pathak. Her title of Chaudhurani would imply that she was a Zamindar, probably a petty one, else she need not have lived in boats for fear of capture.' This passage from W.W. Hunter's *Statistical Account* has been quoted by Jogeshchandra Bagal in his introduction to *Bankim Rachanavali*.
40. This statement and the two others cited in this paragraph are quoted by Jogeshchandra Bagal.
41. It was started by Akshaykumar Maitreya. The manifesto was quoted by Rabindranath in his essay 'Aitihasik Chitra.' See *Rabindra Rachanavali*, XIII, p. 477.

42. See Shantiswaroop Gupta, *Hindi tatha Marathi Upanyason ka Tulanat-mak Adhyayan* (Delhi, 1976), p. 64. Hindi.

43 *Rabindra Rachanavali*, XIII, p. 477.

44. *Bankim Rachanavali*, II, p. 331. This was a review of Rajkrishana Mukhopadhyay's *Prathamsiksha Banglar Itihas (A Primer of Bengal's History)*.

45. Quoted by Sudhir Chandra from *Bharatendu Granthavali*, III, p. 902.

46. *Rabindra Rachanavali*, XIII, p. 477.

47. Ibid.

48. Ibid.

49. Lukacs (first English translation 1962; rptd. Harmondsworth, 1969).

50. Gopal Rai, *Hindi Katha Sahitya Aur Uske Vikas par Pathakon ke Ruchi ka Prabhav* (Patna, 1965), p. 138

51. M.S. Pati and J.K. Nayak, 'History and The Novel in India—the Oriya Historical Novel: A Case Study', *Jadavpur Journal of Comparative Literature*, Vols. 18–19 (1980–1), p. 63.

52. 'The Mogul (Aurangzeb) demanded the hand of the princess of Roonna-gurh, a junior branch of the Marwar house, and sent with the demand (a compliance with which was contemplated as certain) a cortege of two thousand horse to escort the fair to court. But the haughty Rajpootni, either indignant at such precipitation or charmed with the gallantry of the Rana, who had evinced his devotion to the fair by measuring his sword with the head of her house, rejected with disdain the profferred alliance, and justified by brilliant precedents in the romantic history of her nation, she entrusted her cause to the arm of the chief of the Rajpoot race, offering herself as the reward of protection. The family priest (her preceptor) deemed his office honoured by being the messenger of her wishes, and the billet he conveyed is incorporated in the memorial of this reign. "Is the swan to be the mate of the stork: a Rajpootni, pure in blood, to be wife to the monkey-faced barbarian!" concluded with a threat of self destruction if not saved from dishonour. This appeal, with other powerful motives was seized on with avidity by the Rana as a pretext to throw away the scabbard, in order to illustrate the opening of a warfare, in which he determined to put all to the hazard in defence of his country and his faith. The issue was an omen of success to his warlike and superstitious vassalage. With a chosen hand, he rapidly passed at the foot of the Aravulli and appeared before Roopnagurh, cut up the imperial guards, and bore off the prize to his capital. The daring act was applauded by all who bore the name of Rajpoot and his chiefs with joy gathered their retainers around the "red standard" to protect the queen so gallantly achieved.'

   From *Annals and Antiquities of Rajasthan* by James Tod (London: 1929, reprinted New Delhi, 1978), Vol. I, p. 301.

53. See *Bankim Rachanavali*, I, p. 40.

54. Ibid., p. 566.

55. Quoted by Gopal Rai, pp. 309–10.

56. *The Novel in India*, p. 123.

57. Faiz Ahmad Faiz, in his *Mizan Nashirin* (Lahore, 1962). Quoted in

translation by Ralph Russel, *The Novel in India*, p. 132. Aziz in Forster's *A Passage to India* fantasizes about the Mughal past for precisely the same reason: to make present servitude bearable.

58. Gopal Rai, p. 272.
59. *Hindi Sahitya ka Itihas* (Varanasi, 1929), pp. 498–9.
60. *Chandrakanta Santati* (1901), pt xxi, p. 73.
61. *Atharah Upanyas* (New Delhi, 1981), p. 20.

Chapter 4. Women in a New Genre

1. Tzvetan Todorov, 'The Origin of Genres,' *New Literary History*, VIII: 1 (Autumn 1976), p. 164.
2. See Ian Watt, *The Rise of the Novel*, p. 241.
3. Of these four novels, two are easily available in English translation: *Indulekha*, translated into English by W. Dumergue, 1890. (Reprinted by Matrubhumi Printing & Publishing Co., Calicut, 1965). *Umrao Jan Ada*, translated into English by Khushwant Singh and M.A. Husaini for the Unesco Collection of Representative Works, Indian Series. (Reprinted, New Delhi, Sangam Books, 1982). Page references are to these editions. *Indira* was translated into English by J.D. Anderson in 1918 and published by Modern Reviews Office, Calcutta. But I have not been able to trace this edition. The page references to *Indira* are to the Bengali text reproduced in *Bankim Rachanavali*, Vol. I (Calcutta, Sahitya Samsad, 1980). My discussion refers to the revised and enlarged version of 1893. *Pan Lakshyant Kon Gheto* has been issued in different Indian languages by the Sahitya Akademi. My quotations are from the Bengali translation entitled *Kintu Keyi Ba Khabar Rakhe* by Sarojini Kamtanurkar, 1971, although I have also referred to the Marathi edition occasionally. The quotations from *Indira* and *Pan Lakshyant Kon Gheto* are my own literal translations.
4. *Indira*, p. 317.
5. Ibid., p. 335.
6. Ibid., p. 335.
7. Ibid., p. 336.
8. Ibid., p. 335.
9. Ibid., p. 291.
10. *Indulekha*, pp. 367–8.
11. For example, when the *Mahabharata* is read aloud in Bengali, it ends with the following couplet: *Mahabharater katha amrita saman / Kashiram Das kahe shune punyaban* (The story of the *Mahabharata* is like nectar. Kashiram Das narrates it, and those who listen are virtuous.) Incidentally, *Umrao Jan Ada* ends in a similar fashion. See p. 232 of the novel.
12. *Indulekha*, p. xx.
13. Ibid., pp. 137–8.
14. Ibid., p. 200.
15. Ibid., p. xx
16. An English translation was done by G.F. Ward (London, 1908).

17. *Pan Lakshyant Kon Gheto*, p. 377.
18. Ibid., p. 253.
19. Ibid., p. 662.
20. Ian Watt, *Rise of the Novel*, p. 160.
21. *Umrao Jan Ada*, p. 21.
22. Ibid., p. 56.
23. Ibid., p. 188.
24. Ibid., p. 192.
25. Ibid., p. 225.
26. Ibid., p. 167.
27. Harry Levin *The Gates of Horn* (New York, 1963), p. 3.
28. George Eliot, *The Mill on the Floss* (1861; reprinted New York, Signet Classics, 1965), p. 446.

Chapter 5.   The Novelist for All Seasons

1. Wherever it has been necessary to cite the text I have translated from the Centenary Edition of *Sarat Rachanavali* (Bengali Era, 1385–7), published in five volumes by the Sarat Samiti (31, Aswini Dutta Road, Calcutta 29). Reference by volume and page number has been given at the end of each quotation.

2. Saroj Bandopadhyay has suggested that nearly all of Saratchandra's numerous widow heroines are variations on Rabindranath's Binodini. See *Bangla Upanyaser Kalantar* (Calcutta, Sahityashri, 3rd revised and enlarged edn., 1976), p. 250.

3. For example Roma in *Pallisamaj*, Rajlakshmi in *Srikanta*, Savitri in *Charitraheen*, and Neelima in *Sesh Prashna*.

4. Bandopadhyay, p. 261.

5. For a fuller discussion of this in Bengali, see Parthapratim Bandopadhyay, *Upanyaser Samajtatva* (Calcutta, Mousumi Prakashan, 1981), p. 120. Also see Premchand's essay 'Mahajani Sabhyata' for a variation on this theme.

6. Quoted in *Charles Dickens: A Critical Anthology*, ed. Stephen Wall (Harmondsworth, 1970), p. 298.

7. This value-laden term is difficult to translate into English. 'Ill-repute' does not convey its various connotations. The aura around a man who led this life probably originated in a kind of Urdu poetry where the *ruswa* or the *rind* is a familiar figure.

8. Buddhadev Bose, *An Acre of Green Grass* (1948; rpt. Calcutta, 1982), p. 45.

9. Dickens, *Dombey and Son* (Harmondsworth, 1975), p. 121.

10. The English word 'chastity' is applicable to both men and women, but *satitva* or its equivalent in most Indian languages is applicable to women alone. The absence of a term for male constancy is a reflection of Indian social values.

11. Rajat K. Ray, 'Man, Woman and the Novel: The Rise of a New Consciousness in Bengal, 1858–1947', *Indian Economic and Social History Review*, XVI : I (1979), pp. 1–32.

12.   Sudhin Ghose did not write in Bengali. His four English novels were inter-connected—like *Srikanta*—and might well have been influenced by Saratchandra's tetralogy. The unnamed hero leads a rootless wandering life. Ghose's mode of narration is however far more fanciful and allegorical than Saratchandra's. His four novels are *And Gazelles Leaping* (1949), *Cradle of the Clouds* (1951), *The Vermilion Boat* (1953), and *The Flame of the Forest* (1955), all published by Michael Joseph, London.

13.   *The Flame of the Forest* (London, 1955), p. 221.

14.   Vasant Bapat, 'Saratchandra and the Marathi Novel', *Indian Literature* (July-August 1979), p. 11.

15.   Jainendra Kumar, *Wey aur Wey* (New Delhi, 1977), p. 87. Hindi.

Chapter 6.   **Pather Panchali**

1.   George Lukacs, *Theory of the Novel*, trans. Anna Bostock (Cambrige, Mass., 1971), p. 97.

2.   Ibid., p. 126.

3.   Robert Scholes, 'An Approach through Genre' in *Towards a Poetics of Fiction*, ed. Mark Spilka (Bloomington, 1977), p. 48.

4.   Two different English translations of the first volume have been published. These are: *The Song of the Road*, trans. T.W. Clark and Tarapada Mukherjee (London, George Allen & Unwin, 1968) and *Pather Panchali*, in three parts, trans. Monika Varma (Calcutta, Writers Workshop, 1973). The first is scrupulously faithful to the original but leaves out the last six chapters. The second is a complete but somewhat free rendering. The second volume of the original, *Aparajito*, has not yet been translated into English.

5.   *The Song of the Road*, pp. 303–4.

6.   Edwin Gerow, 'Three Bengali Novels' in *The Literatures of India: An Introduction*, ed. Edward Dimock and others (Chicago, University of Chicago Press, 1974), p. 236.

7.   For a fuller discussion of the different versions in English, see Sujit Mukherjee, *Translation as Discovery* (New Delhi, Allied Publishers, 1981), Chapter 7.

8.   In his 'Afterword' to *Pather Panchali*, Part III, trans. Monika Varma, Naresh Guha has given a list of books read by Opu as mentioned in the novel and discussed the importance of this eclectic reading in Opu's imaginative growth.

9.   For a detailed contrastive study of these two novels, see Meenakshi Mukherjee, 'The House and the Road: Two Modes of Autobiographical Fiction' in *Commonwealth Literature: Problems of Response,* ed. C.D. Narasimhaiah (Madras, 1981), pp. 148–64.

10.   My translation from the Mitra and Ghosh Bengali paperback reprint of 1972. This extract is from page 70.

11.   *The Song of the Road*, pp. 75–6.

12.   See the essay 'Calcutta' in his *The World of Twilight* (Calcutta, Oxford University Press, 1970), pp. 77–8.

13.   See Edwin Gerow.

14.   My translation, p. 165.

15. For an elaboration of this view see 'Decolonising the Mind' , a supplement in *South* (January 1983).
16. Lukacs, p. 130.
17. About the response of readers to this novel I should like to record an empirical observation. From the experience of teaching this novel at an American university as well as at an Indian university—neither experience being long enough to support generalization—I have found two kinds of reactions from students. American students and Indian students with a predominantly western cultural orientation tend to react sharply to the socioeconomic aspects of life as depicted in the novel; they appear to be struck forcefully by the poverty and the deprivation that lead to Durga's death and Opu's displacement. Whereas the more indigenous Indian students— be they from Andhra Pradesh or Bengal or Kerala—react more deeply to the human aspects, noticing not so much the poverty, which in any case is an ubiquitous Indian presence, but Opu's sense of joy and wonder at life.
18. Naresh Guha, p. iii.
19. Nabaneeta Deb Sen, 'Two Cases of Conscience and Alienation', *Jadavpur Journal of Comparative Literature*, XI (1973), pp. 107–23.

Chapter 7.   Godan

1. Of the writings of Premchand mentioned in this paragraph, *Godan* is available in two different English translations (see note 7 below) and 'Kafan' is included in English translation in the anthology titled *Twenty Four Stories by Premchand*, ed. Nandini Nopany and P. Lal. The presidential address delivered at the inaugural meeting of AIPWA is not available in English translation. The Hindi version is included in the volume of essays entitled *Sahitya ka Uddeshya* (Allahabad, Hans Prakashan, 1967). The essay 'Mahajani Sabhyata' ('Capitalist Civilization') is included in the Hindi volume *Mangalsutra wa anya rachanaen* (Allahabad, Saraswati Press, no date). No English translation is available.
2. *Sahitya ka Uddeshya*, p. 17.
3. Engels' letter to Margaret Harkness in April 1888, quoted by George Steiner in 'Marxism and the Literary Critic', in *Language and Silence* (1958; rpt. Harmondsworth, 1969), p. 271.
4. Since film is a more pervasive medium today than fiction, many people may be familiar with this story through its film version made by Mrinal Sen in Telugu *(Oka Oori Katha)*. However, the intensity of the short story depends upon its brevity (it is only nine pages long) while the film achieves another kind of effect by expansion.
5. Anton Pavlovich Chekhov, letter to A.S. Suvorin, 27 Oct. 1888, in *Selected Letters*, trans. S. Lederov, ed. L. Hellman (1955), p. 57.
6. '*Godan* is the epic of Indian life in the form of a novel'—Gopal Krishna Kaul. From *Godan: Mulyankan aur Mulyankan*, ed. Indranath Madan (Allahabad, 1971).
7. *Godan*, translated by P. Lal and Jai Ratan (Bombay, 1957) and *The Gift of a Cow*, translated by Gordon Rodarmel (Bloomington, 1968). All page references to *Godan* in this volume are to the first translation.

8. Nandadulare Bajpeyi considers the two sections as separate as two flats in the same house. He does not see any justification for the urban strand. See 'Godan : ek sahityik vivechan' in *Godan: Mulyankan aur Mulyankan*. Ramdeen Gupta sees in this novel a transition from an enfeebled feudalism which cannot stand on its own, to a purely capitalistic society. See his essay in Madan.

9. The terms 'story' and 'discourse' are being used here in their generally accepted meaning in studies in narratology (by Roland Barthes, Tzvetan Todorov, Gerard Genette and Seymour Chatman, for example). The 'what' of the narrative is the story (*histoire*) and the 'how' of the narrative is the discourse (*discours*). For further explanation of the term see Seymour Chatman, *Story and Discourse: Narrative Structure in Fiction and Film*, Introduction (Cornell, 1978).

10. Mimesis and diagesis are distinctions made by Plato and the terms have been revived by Gerard Genette in 'Frontiers du recit', *Communications: 8* (1966). The difference is essentially between direct presentation and mediated narration. The terms 'representational' and 'illustrative' have been explained thus:

    The connection between the fictional world and the real can be either representational or illustrative. The images in a narrative may strike us at once as an attempt to create a replica of actuality . . . or they may strike us as an attempt merely to remind us of an aspect of reality rather than convey a total and convincing impression of the real world to us . . . .

    Robert Scholes and Robert Kellog, *The Nature of Narrative* (Oxford, 1971 reprint), p. 84.

11. Sumitranandan Pant, 'Adhunika,' *Gramya* (Allahabad, 1977), p. 83.

12. 'Mahajani Sabhyata,' *Mangalsutra wa Anya Rachanaen* (Allahabad, n.d.), p. 365.

13. 'If the feudal lord quenched his thirst with his enemy's blood, he also risked his own life for a friend or a benefactor. If the emperor's will was law and he would not tolerate disobedience, he could also be kind to his subjects and dispense justice impartially . . . The king did not think of his people merely as the fuel for his fire of acquisition; he also shared their joys and sorrows and was a patron of talents.

    But in this civilization based on capitalism, money is the sole motivation for work . . .' 'Mahajani Sabhyata', *Mangalsutra wa Anya Rachanaen*, p. 365.

14. Premchand once wrote in a letter that in his ideal of womanhood the qualities of service, sacrifice and purity (*tyag, seva aur pavitrata*) converged. See Indranath Madan, *Premchand, Ek Vivechan*, p. 247.

15. Govindi, incidentally, was called Kamini when she was first introduced in Chapter Six. Not only did Premchand change her name later, he also changed her entire character. As Kamini she was petulant, constantly complaining of headaches, and was said to be a writer of doggerel. As Govindi she is not only put on a pedestal but her poems also improve. Mehta praises them in a later chapter.

16. Rabindranath Tagore, *Gora* (1907), p. 311.

17.   See 'The Search for an Identity', *Vagartha: 15* (October 1976), pp. 1–16.
18.   Ibid., p. 8.
19.   See M.N. Srinivas, *Social Change in Modern India*, (1977 reprint), p. 136.
20.   Max Weber in *The Protestant Ethic and the Spirit of Capitalism* (1930) argues that a persistent feature of economic individualism is that the furtherance and promotion of economic motive logically entails a devaluation of other modes of thought, feeling and action—mainly the various forms of traditional group relationships. The family, filial duties, the village, everything is weakened and placed against the competing claims of individual achievement.
21.   See Ian Watt, *Rise of the Novel*, p. 65.
22.   J.S. Mbiti, *African Religions and Philosophy* (London, 1969), p. 225.
23.   'Firstly, in the fact that labour is *external* to the worker, i.e., it does not belong to his essential being, in the fact that he therefore does not affirm himself in his work, but negates himself in it, that he does not feel content, but unhappy in it, that he develops no free physical and mental energy but mortifies his body and ruins his mind. Therefore the worker feels himself only outside his work, while in his work he feels outside himself. He is at home when he is not working and when he works he is not at home. His work, therefore, is not voluntary but coerced; it is *forced* labour. It is, therefore, not the satisfaction of needs but only a *means* for satisfying needs external to it. Its alien character emerges clearly in the fact that as soon as there is no physical or other compulsion, labour is avoided like the plague. External labour, in which man alienates himself, is labour of self-sacrifice, of mortification. Finally, the external character of labour for the worker appears in the fact that it is not his own but somebody else's, that activity is not his own activity. It belongs to another, it is the loss of his self.

The result, therefore, is that man (the worker) no longer feels himself acting freely except in his animal functions, eating, drinking and procreating, or at most in his dwelling, ornaments, etc., while in his human functions he feels more and more like an animal. What is animal becomes human and what is human becomes animal.

Drinking, eating and procreating are admittedly also genuinely human functions. But in their abstraction, which separates them from the remaining range of human functions and turns them into sole and ultimate ends, they are animal.'

Karl Marx, 'Economic and Philosphical Manuscripts of 1844', Marx and Engels, *Complete Works*, Vol. 3 (Moscow, 1975), p. 274.
24.   *Sarat Rachanavali*, Vol. I, p. 221.
25.   Being a labourer is for Hori a matter of humiliation—as he says very clearly on p. 17:

Look at us for instance. What do we get out of our land? The return does not even equal an anna's wage per head. A servant with a salary of ten rupees per month is better off than us. But we do not give up our land and go in for a job. Do we? Our prestige gets in the way.

26. Gayatri Spivak, 'Version of a Colossus,' *Vagartha: 24* (January 1979), p. 55.
27. 'Adarshvadi yatharthavad'—his own term in the AIPWA address.
28. Premchand has been likened to Hardy in terms of the gradual change in his attitude towards the rural community: 'The dissolving hope and growing despair, though paced differently, follow the same direction, rendering the ambivalence of each writer increasingly complex. As the triumphs grow fewer and less convincing, Hardy and Premchand come to terms with their recognition of rural failure. They see that rural prople, even the finest, are culpable because they allow themselves to be exploited; that the country has lost its beauty; that what remains is coarse and ugly as in *Jude* or ossified and cruel as in *Godan*'. See Zarina Manawar Hock, 'Force and Counterforce: The Rural-Urban Encounter in the Novels of Thomas Hardy and Premchand', unpublished dissertation.
29. Georg Lukacs, 'Tolstoy and the Development of Realism' in *Marxists on Literature*, ed. David Craig, p. 259.

Chapter 8. Samskara

1. S. Nagarajan holds that 'this criticism is not extra-literary since it involves the character's mimetic mode'. See *'Samskara,' Indian Writing Today: 17* (July—Sept. 1971), p. 123.
2. In *India: A Wounded Civilization* (London, 1976).
3. Ibid.
4. The comments on the novel in Kannada are not available to me. Among the interpretations I have in English so far, the following seem to me the most important:
   —S. Nagarajan, *'Samskara'*.
   —Madhu Prasher, 'Another View of Samskara,' *Vagartha: 15* (October 1976), pp. 25–31.
   —Margaret Nightingale, 'Anantha Murthy's *Samskara:* an Indian Journey to the End of Night,' *New Literature Review*, No. 4, pp. 51–4.
   —R.K. Kaul, *'Samskara'*; a paper read at a seminar in the University of Rajasthan in 1981.
5. Lukacs, *Theory of the Novel*, p. 66.
6. Northrop Frye in *Encyclopaedia of Poetry and Poetics*, ed. Alex Preminger (Princeton, 1965), p. 12.
7. Rajat K. Ray 'Peasant and Landless Untouchables in the Fiction of the Gandhian Age', a paper presented to a seminar at the Nehru Memorial Museum and Library in February 1983.
8. R.K. Kaul in a paper read at a seminar in the University of Rajasthan said that the challenge to the brahmans comes not from Naranappa alone; 'The challenge comes from a force that is powerful but evil.'
9. Jaya and Vijaya were the gate keepers of Vishnu who were cursed by the angry sages. Later, after being placated, the sages conceded they could return to God in two different ways: either by worshipping Him as a devotee or by being His enemy. The first would take them seven lives to reach God and the second only three. In the

next three lives Jaya was Hiranyaksha, Ravana and Shishupala respectively, and Vijaya was Hiranjyakashipu, Kumbhakarana and Dantabakra. They were united with Vishnu by being killed by him. My source for this is *Pouranik Abhidhan*, ed. Sudhir Chandra Sankar (Calcutta, 1953), p. 185.

10. Later in the novel there is suggestion in Putta's prattle that Padmavati too was a woman Naranappa had enjoyed, and the Acharya's irresistible attraction for her suggests another link between the two antagonists.

11. '*Papoham papakaramaham*'.

12. See S. Nagarajan, *Indian Writing Today*, July-Sept. 1971, p. 122.

13. Ibid

14. Anton Chekhov, letter to A.S. Suvorin, 27 October 1888, in *Selected Letters*, trans. S. Lederov, ed. Lillian Hellman (1955), p. 57.

BIBLIOGRAPHY

A. Secondary Sources

Bandopadhyay, Parthapratim. *Upanyaser Samajtatva* (Sociology of the Novel. Bengali). Calcutta, 1981.

Bandopadhyay, Saroj. *Bangla Upanyase Kalantar* (Changing Epochs in the Bengali Novel. Bengali). Calcutta, 1961. Revised and enlarged third edition, 1976.

Bandopadhyay, Srikumar. *Bangashitye Upanyaser Dhara* (Tradition of the Bengali Novel. Bengali). Calcutta, 1939. Third edition, 1956.

Bhate, G.C. *History of Marathi Literature.* Pune, 1939.

Bose, Buddhadev. *An Acre of Green Grass: A Review of Modern Bengali Literature.* Calcutta, 1948. Reprinted, 1982.

Bankim Rachanavali (Collected Works of Bankimchandra Chattopadhyay. Bengali), ed. Jogeshchandra Bagal. 2 vols. Calcutta, 1980. Ninth edition.

Chaitanya, Krishna (K.K. Nair). *History of Malayalam Literature.* New Delhi, 1971.

Chatman, Seymour. *Story and Discourse: Narrative Structure in Fiction and Film.* Ithaca, 1978

Chekhov, Anton Pavlovich. *Selected Letters.*Trans. L. Lederov and ed. L. Hellman. New York, 1955.

Clark, T.W. (ed.). *The Novel in India.* London, 1970.

Craig, David (ed.). *Marxists on Literature.* Penguin Books, 1975.

Datta, Bijitkumar. *Bangla Sahitye Aitihasik Upanyas* (The Historical Novel in Bengali Literature. Bengali). Calcutta, 1963.

Deshpande, Kusumavati. *Marathi Kadambari: Pahile Shatak* (The First Century of the Marathi Novel. Marathi). 2 vols. Pune, 1953-4.

Dimock, Edward (ed.). *Literature and History.* Ann Arbor, Michigan, 1967.

——and others (ed.). *The Literatures of India: An Introduction.* Chicago, 1974.

Eliade, Mircea. *Myth and Reality.* New York, 1963. Harper Torchbook edition, 1968.

Frye, Northrop. *Anatomy of Criticism*. Princeton, 1957. Princeton Paperback edition, 1971.

Georgakas, Dan and Lubenstein, Lenny (ed.). *The Cineaste Interviews on the Art and Politics of the Cinema*. Chicago, 1983.

Gupta, Pramila. *Premchand aur Harinarain Apte ka Tulanatmak Adhyayan* (A Comparative Study of Premchand and Harinarain Apte. Hindi). Delhi, 1970.

Gupta, Shantiswarup. *Hindi tatha Marathi Upanyas ka Tulanatmak Adhyayan, 1900–50* (A Comparative Study of the Hindi Novel and the Marathi Novel, 1900–50. Hindi). New Delhi, 1976.

Jussawalla, Adil. *New Writing in India*. Penguin Books, 1974.

Kakar, Sudhir (ed.). *Identity and Adulthood*. New Delhi, 1979.

Karve, Irawati. *Yuganta: End of an Epoch*. Translated from her Marathi work, *Yuganta*. Sangam Books edition, 1973.

Levin, Harry. *The Gates of Horn*. New York, 1963.

Lukacs, Georg. *The Historical Novel*. Trans. Hannah and Stanley Richard. Boston, 1963. Penguin Books, 1969.

——*Studies in European Realism*. Trans. Edith Bone. Reprinted London, 1978.

——*Theory of the Novel*. Trans. Anna Bostock. Cambridge, Mass., 1971.

Madan, Indranath (ed.). *Godan: Mulyankan aur Mulyankan* (Godan: Evaluation and Evaluation. Hindi). Allahabad, 1971.

——*Premchand—Ek Vivechan* (A Consideration of Premchand. Hindi). Delhi, 1968.

Marx, Karl. 'Economic and Philosophic Manuscripts of 1844', in *The Collected Works of Marx and Engels*, Vol. III. Moscow, 1975.

Mbiti, J.S. *African Religions and Philosophy*. London, 1979.

Mukherjee, Meenakshi (ed.). *Considerations*. New Delhi, 1977.

Mukherjee, Sujit. *Translation as Discovery & Other Essays*. New Delhi, 1981.

Munshi, K.M. *Gujarat and Its Literature*. Bombay, 1935.

Naipaul, V.S. *India: A Wounded Civilization*. London, 1976.

Narasimhaiah, C.D. (ed.). *Commonwealth Literature: Problems of Response.* Madras, 1981.

O'Malley, L.S.S. (ed.). *Modern India and the West.* London, 1941. Reprinted 1968.

Premchand, Munshi. *Sahitya ka Uddeshya* (The Purpose of Literature. Hindi). Allahabad, 1967.

——*Mangalsutra wa Anya Rachnaen* (Mangalsutra and Other Essays: Hindi). Allahabad, 1948.

Rabindra Rachanavali (Complete Works of Rabindranath Tagore: Bengali). Centenary Edition. Calcutta, 1962.

Rai, Amrit. *Premchand:A Life.* Trans. Harish Trivedi, New Delhi, 1982.

Rai, Gopal. *Hindi Kathasahitya aur Uske Vikas par Pathakon ki Ruchi ka Prabhav* (The Influence of Readers' Taste on the Development of Hindi Fiction: Hindi). Patna, 1965.

Ronaldshay, Lord. *The Heart of Aryavarta.* London, 1929. Indian reprint New Delhi, 1980.

Said, Edward. *Beginnings: Intentions and Method.* New York, 1975.

Sarat Rachanavali, 5 Vols. (Complete Works of Saratchandra Chattopadhyay. Bengali). Centenary Edition. Calcutta, 1976.

Scholes, Robert and Kellogg, Robert. *The Nature of Narrative.* New York, 1966. Reprinted 1971.

Sen, Sukumar. *History of Bengali Literature.* New Delhi, 1960. Revised edition, 1971.

Shukla, Ramchandra. *Hindi Sahitya ka Itihas* (History of Hindi Literature: Hindi). Varanasi, 1929.

Srinivas, M.N. *Social Change in Modern India.* Berkeley, 1962.

Steiner, George. *Language and Silence.* New York, 1967.

Spilka, Mark (ed.). *Towards a Poetics of Fiction.* Indiana, 1977.

Tod, James. *Annals and Antiquities of Rajasthan.* London 1929. Indian reprint, New Delhi, 1979.

Varshneya, Lakshmisagar. *Adhunik Hindi Sahitya ki Bhumika.* (Introduction to Modern Hindi Literature. Hindi). Allahabad, 1952.

Watt, Ian. *The Rise of the English Novel.* Penguin Books, 1983.

Weber, Max. *Protestant Ethic and the Spirit of Capitalism.* London, 1930.

Yadav, Rajendra. *Atharah Upanyas* (Eighteen Novels; Hindi). New Delhi, 1981.

B. List of Indian Novels in English Translation (mentioned in this book)[1]

'Agyeya' (S.H. Vatsyayan). *Apne Apne Ajnabi* (Hindi: 1961). Trans. author, *To Each His Stranger* (Delhi, Orient Paperbacks, 1967).

Anantha Murthy, U.R. *Samskara* (Kannada: 1965). Trans. A.K. Ramanujan (Delhi, Oxford University Press, 1976).

Bandopadhyay (Banerji), Bibhutibhusan. *Pather Panchali* (Bengali: 1929). Trans. T.W. Clark and Tarapada Mukherjee, *The Song of the Road* (London, Allen & Unwin, 1968). Trans. Monika Varma, *Pather Panchali*, in three parts (Calcutta, Writers Workshop, 1973).

Bandopadhyay (Banerji), Manik. *Putul Nacher Itikatha* (Bengali: 1936). Trans. Sachindralal Ghosh, *The Puppet's Tale* (Delhi, Sahitya Akademi, 1968; reprinted 1978).

Bandopadhyay (Banerji), Tarashankar. *Arogyaniketan* (Bengali: 1952). Trans. Enakshi Chatterjee, *Arogyaniketan* (New Delhi, Arnold Heinemann, 1977).

Bhattacharya, Lokenath. *Babughater Kumari Machh* (Bengali: 1970). Trans. Meenakshi Mukherjee, *The Virgin Fish of Babughat* (New Delhi, Arnold Heinemann, 1974).

Karanth, K. Sivaram. *Marali Mannige* (Kannada: 1940). Trans. A.N. Murthy Rao, *The Whispering Earth* (Delhi, Vikas Publishing House, 1974).

——*Chomana Dudi* (Kannada: 1933). Translator not mentioned (Delhi, 1978).

Menon, O. Chandu. *Indulekha* (Malayalam: 1889). Trans. W. Dumergue, *Indulekha* (1890; reprinted Calicut, Matrubhumi Publishing & Printing, 1965).

Nagarkar, Kiran. *Sat Sakkam Trechalis* (Marathi: 1974). Trans. Shubha Slee, *Seven Sixes Are Forty-Three* (Delhi, Vikas Publishing House. 1981).

---

[1] For a fuller listing of novels available in English translation, see the bibliography in Sujit Mukherjee, *Translation as Discovery & Other Essays* (New Delhi, Allied Publishers, 1981).

Premchand, Munshi. *Godan* (Hindi: 1936). Trans. P. Lal and Jai Ratan, *Godan* (Bombay, Jaico Books, 1957). Trans. Gordon Rodarmel, *The Gift of a Cow* (Bloomington, Indiana University Press, 1968).

Ruswa, Mirza Hadi. *Umrao Jan Ada* (Urdu: 1901). Trans. Khushwant Singh and M.A. Husain, *Umrao Jan Ada* (Calcutta, Orient Longman, 1961; reprinted Delhi, Sangam Books, 1982).

Tagore (Thakur), Rabindranath. *Gora* (Bengali: 1910). Trans. probably W.W. Pearson, *Gora* (London, Macmillan, 1924; has been reissued in Macmillan Pocket Tagore edition, subsequently).

Vaid, Krishna Baldev. *Uska Bachpan* (Hindi: 1957). Trans. author, *Steps in Darkness* (New York, Orion Press, 1962; reprinted Delhi, Orient Paperbacks, 1968).

INDEX